Lapologang Magole, Alyne Delaney (Eds.)

Partnerships and Power Games

Defragmenting African Resource Management

edited by

Associate Professor Dr. Mafaniso Hara
(University of the Western Cape)

and

Prof. Dr. Tobias Haller
(Universität Bern)

PLAAS
Institute for Poverty, Land and Agrarian Studies

A publication of the
Institute for Poverty, Land and Agrarian Studies (PLAAS)
University of the Western Cape
South Africa

Volume 4

ACP S&T PROGRAMME

LIT

Partnerships and Power Games

Natural Resources Governance and Management in the Okavango Delta

edited by

Lapologang Magole, Alyne Delaney

LIT

Photo credits: L-R (top-bottom):
1.–4., Alyne Delaney
1. Cattle grazing outside Thubu Village. Livestock are an important source of livelihood.
2. Ecotourism sites have access to riverfronts and in many places block communities from accessing natural resources.
3. Traditional handmade dugout canoes on the river bank. Traditional skills often go hand-in-hand with natural resource use and management.
4. Children on donkeys, outside of the Tsodilo UNESCO World Heritage Site area, which is now managed by a community trust as an ecotourism venture.

Maps and illustrations by Masego Dhliwayo

The editors would like to thank the European Union (project AFS/2009/219017) through the African Caribbean and Pacific (ACP) Science and Technology Programme (Europe Aid/127860/D/ACT/ACP) for the funding that made this book possible. The contents of this book though are our sole responsibility and do not reflect the views or position of the European Union.

This book is printed on acid-free paper.

Bibliographic information published by the Deutsche Nationalbibliothek
The Deutsche Nationalbibliothek lists this publication in the Deutsche Nationalbibliografie; detailed bibliographic data are available on the Internet at http://dnb.d-nb.de.

ISBN 978-3-643-90685-4

A catalogue record for this book is available from the British Library

Klosbachstr. 107
CH-8032 Zürich
Tel. +41 (0) 44-251 75 05
E-Mail: zuerich@lit-verlag.ch http://www.lit-verlag.ch
Distribution:
In the UK: Global Book Marketing, e-mail: mo@centralbooks.com
In North America: International Specialized Book Services, e-mail: orders@isbs.com
In Germany: LIT Verlag Fresnostr. 2, D-48159 Münster
Tel. +49 (0) 2 51-620 32 22, Fax +49 (0) 2 51-922 60 99, e-mail: vertrieb@lit-verlag.de
In Austria: Medienlogistik Pichler-ÖBZ, e-mail: mlo@medien-logistik.at
e-books are available at www.litwebshop.de

Table of Contents

List of Tables

List of Figures

Acronyms

AWIRU	African Water Issues Research Unit
BCCT	Bukhakwe Conservation and Cultural Trust
BNES	Botswana National Ecotourism Strategy
CBO	Community-Based Organisation
CBRM	Community-Based Resource Management
CBNRB	Community-Based Natural Resource Broker
CBNRM	Community-Based Natural Resource Management
CBNRMP	Community-Based Natural Resource Management Project
CBPP	Contagious Bovine Pleuropneumonia (Cattle-Lung Disease)
CHA	Controlled Hunting Area
DAHP	Department of Animal Health and Production
DARMA	Defragmenting African Resource Management
DEA	Department of Environmental Affairs
DFRR	Department of Forestry and Range Resources
DLUPU	District Land Use Planning Unit
DWA	Department of Water Affairs
DWNP	Department of Wildlife and National Parks
EBFM	Ecosystem Based Fishery Management
FMD	Foot and Mouth Desease
HOORC	Harry Oppenheimer Okavango ResearchCentre
INRM	Integrated Natural Resources Management
ISPAAD	Integrated Support Programme for Arable Agricultural Development
IWRM	Integrated Water Resource Management
JMC	Joint Management Committee
LIMID	Livestock Management Infrastructure Development
MEWT	Ministry of Environment, Wildlife and Tourism
MMEWR	Ministry of Minerals, Energy and Water Resources
MOA	Ministry of Agriculture
MOMS	Management Oriented Monitoring System
NAMPAADD	National Master Plan for the Arable Agriculture and Dairy Development
NDSS	Ngamiland District Settlement Strategy
NMPWWS	Botswana National Master Plan for Waste Water and Sanitation

NWMPR	National Water Master Plan Review
OCT	Okavango Community Trust
ODIS	Okavango Delta Information System
ODMP	Okavango Delta Management Plan
OKACOM	Okavango River Basin Water Commission
ORI	Okavango Research Institute
ORPMP	Okavango River Panhandle Management Plan
SLM	Sustainable Land Management
SOIWDP	Southern Okavango Integrated Water Development Project
TGLP	Tribal Grazing Land Policy
VDC	Village Development Committee
VET	Village Extension Team
WMA	Wildlife Management Areas

DARMA Book Series Foreword

In Southern Africa the ‘commons’ have great livelihoods and biodiversity value, and they function as ‘common pool resources’ that are owned, accessed and used by multiple stakeholders in the region. However, as in many other parts of the world, management of the Southern African commons is fragmented. This fragmentation occurs in many ways and at different levels. For instance, authority is located in different government line agencies and technical expertise is divided among government agencies, non-governmental organisations (NGOs), universities and donors. In addition, actions take place at different geographic scales, and management networks are usually intra-sectoral (for example those in fisheries only know and network with others in fisheries). Funding for research and management also comes from multiple sources including foreign and international donors; and in some cases, commons transcend international borders, requiring international integration. Fragmentation in policy, legislation, knowledge base, authority and responsibility clearly makes a holistic approach to commons management difficult.

User stakeholders, especially the poor, usually draw upon the various resources (land, forests, fish, grass, water, wildlife) for composite benefits and livelihoods. If managed holistically, these commons can remain highly productive and sustainable. However, commons are in most instances under pressure due to their economic value and uncoordinated management. Overexploitation and degradation occur particularly where ecosystem-based and ecosystem-wide institutional arrangements do not exist to overcome problems of subtractability that arise from open access.

In recent years, there have been increasing attempts to manage commons holistically as a result of the realisation and appreciation of the interrelatedness of both the ecological and socio-economic dynamics of natural resources – the so-called ‘ecosystems approach’. This approach calls for partnerships between state agencies, research institutions, NGOs, donors, traditional leaders, local communities and all other stakeholders. Effective links and collaborative partnerships among all stakeholders are critical for ensuring that the available scientific knowledge, indigenous knowledge, user knowledge and institutions are optimised. This book is an effort at demonstrating such a multi-disciplinary and inter-disciplinary approach in management of the Okavango Delta environs. It is based on contributions from various experts and users with long-term experience and knowledge of the area. The book reviews and evaluates policies and management approaches for the Okavango Delta environs, describes how these are fragmented and proposes a framework for defragmenting management.

The Okavango Delta case study is one of seven cases under the Defragmenting African Resource Management project (DARMA) in Southern Africa, which aimed to build networks and research frameworks that could integrate resource management. The specific objective of DARMA was to increase national capacity in five African partner countries (Botswana, Malawi, South Africa, Zambia, Zimbabwe) for integrated management of commons with the overall broader objective being enhancement of biodiversity of exploited commons while alleviating poverty at the same time. The main strategy for achieving these objectives was the development of scientific networks and bases that can combine different relevant natural and social science disciplines in a way that will be of practical use for both biodiversity conservation and sustainable development. The final outputs of the DARMA project are a series of books and journal articles that will serve as a valuable collection for environmental policy makers and will, hopefully, serve as catalysts in future debates on defragmentation of natural resources management in Southern Africa.

The preparation of this series of books has been made possible with European Union funding (Project AFS/2009/219017) through the African Caribbean and Pacific (ACP) Science and Technology Programme (Europe Aid/127860/D/ACT/ACP). The contents of these books are the sole responsibility of the authors. They do not reflect the views and position of the European Union. We thank all the stakeholders and participants who generously donated their time and energy to making the DARMA project possible.

Development of the research frameworks for analysing fragmentation and defragmentation for the various DARMA case studies were developed jointly by the partners and associate partners. The DARMA partners and associate partners would particularly like to remember and acknowledge the late Professor Douglas Clyde Kongshøj Wilson (Innovative Fisheries Management, Aalborg University, Department of Development and Planning), an associate on the project, for all his contributions to the project.

Associate Professor Mafaniso Hara

DARMA Project Coordinator

Institute for Poverty, Land and Agrarian Studies

Faculty of Economic and Management Sciences

University of the Western Cape

South Africa

Contributors

Bontekanye Botumile is currently a Mphil (Natural Resource Management) student at the Okavango Research Institute – University of Botswana. She is a Bessie Head literature award winner and has published four children's books based on Tswana culture. Her background includes eighteen years' work experience in the tourism industry and in Community Based Natural Resource Management) programs. Her area of research interest and focus is traditional land resources management.

Lin Cassidy works as an environmental consultant focusing on remote sensing, land change science and political ecology. She is based in her home town of Maun, Botswana, where her work addresses sustainable development issues at all scales: from the Okavango Delta, through Botswana as a whole, to the southern African region, and beyond. Lin has a PhD in Geography, and an MS in Interdisciplinary Ecology, both from the University of Florida. She has recently worked as a research scholar at the University of Botswana's Okavango Research Institute. Lin's work and research have taken her to many parts of the world, including the Caribbean and south-east Asia. Her general interests are in human-environment interactions, social-ecological systems, and integrated conservation and development.

Alyne Delaney (PhD anthropology, University of Pittsburgh, USA) is Head of Research and an Associate Professor in Aalborg University's Department of Development and Planning's Innovative Fisheries Management (IFM) research centre where her research focuses on social organization, sustainability and resilience. Her interest in the commons began while studying resource rights and fishing ground access among small-scale fishers in Japan and she has since worked in Southern Africa, SE Asia, Europe and Greenland. In addition to research and teaching, she is currently the chief editor of *The Commons Digest*, the newsletter of the International Association for the Study of the Commons.

Rachel DeMotts (MA and PhD, Political Science, University of Wisconsin-Madison, 2000, 2005) is Mellon Associate Professor of Global Environmental Politics at the University of Puget Sound in Tacoma, Washington. Her research and teaching engage issues of community-based conservation, gender and natural resources, trans-boundary conservation, and tourism in

Southern Africa. She has been researching in southern Africa since 2006, predominantly in community-based conservation projects in Namibia and Botswana with a preferred focus on women affairs and is a former research fellow at the University of Botswana's Okavango Research Institute.

Phemo K. Kgomotso has a Doctorate of Philosophy (DPhil) in Development Studies from the Institute of Development Studies, University of Sussex, UK. Her research interests are natural resource governance and the discourses of access and control over contested landscapes and she has focused on the Okavango Delta resources politics for several years. She currently works as a Technical Advisor on ecosystems and biodiversity as part of the Global Environmental Finance Unit of the United Nations Development Programme, which supports several African countries to design and implement programmes in line with their obligations under the multilateral environmental agreements/UN environmental conventions. She is currently based in Addis Ababa, Ethiopia.

Innocent L. Magole holds a Masters Degree in Wildlife Management from Reading University (UK). He is studying for a PhD in Tourism Management at North West University, South Africa. Innocent has over twenty years' experience in CBNRM with special emphasis on community-based tourism and enterprise development. He is currently Managing Director for SESU Holdings, a consulting firm that undertakes consultancy work on environmental management, conservation, tourism, livelihoods and offers training on these areas. His research interest and focus is on natural resources access and livelihoods.

Lapologang Magole is a Senior Research Scholar at the University of Botswana's Okavango Research Institute, Botswana working in the area of natural resources governance. She holds a PhD in Development Studies (Environmental Policy Analysis) from the University of East Anglia, UK. Her area of research interest and focus is natural resources governance in general and common property resources management in particular. Through her research and application (development work) she has interacted with resource users at grassroots level, donors, policy makers and implementers and has come to appreciate multiple interests and aims with regards to natural resource management and use.

Wellington Riach Liamba Masamba was awarded both a PhD and an MSc in Chemistry from the University of Florida in Gainesville. He is currently Associate Professor in Chemistry and the Acting Director of the Okavango Research Institute. He has taught courses in analytical, inorganic, physical and environmental chemistry, waste management, environmental toxicology and water quality. His research interests include water/wastewater quality, nutrient cycling and plant growth promotion, methane and nitrous oxide fluxes and the determination of essential oils from plants. For over a decade he has collaborated with other water experts and practitioners from Southern and Eastern Africa, through the Waternet platform, to develop the research and development agenda of water resource management.

Joseph E. Mbaiwa is a Social Scientist with training in Environmental Sciences, Natural Resource Management and Tourism Sciences. He holds a PhD in Park, Recreation and Tourism Sciences from Texas A and M University (USA). He is a Professor of Tourism Studies at the Okavango Research Institute, University of Botswana where he is currently the Acting Deputy Director in-charge of Research Management. His research focus is on tourism development, rural livelihoods and biodiversity conservation. Prof Mbaiwa has accumulated wide research experience in tourism development, rural livelihoods, CBNRM, environmental management, natural resource use and biodiversity conservation in Botswana with particular reference to the Okavango Delta.

Mercy H. Moemedi holds a National Diploma in nature conservation from Tshwane University of Technology, South Africa. She has for several years been involved and gained vast experience in environmental assessment and education research projects which involved academics, policy makers and communities in a process of co-learning. Her area of interest is community wildlife resources management and sustainable livelihoods.

Ketlhatlogile Mosepele is an Associate Professor at the Okavango Research Institute, University of Botswana. He holds an Mphil and is a PhD candidate in Fisheries Biology and Management at the University of Bergen, Norway. Ketlhatlogile is currently both the Okavango Research Institute Acting Deputy Director Research Services and Training and the Training Coordinator. His key research areas include assessing the role that fisheries management plays in sustainable development; assessing the relationship between floodplain fish and fishery dynamics with environmental variability, and how this is integrated into management; the study of freshwater ecosystem dynamics, fish biology and aquatic ecology.

Moseki Ronald Motsholapheko is a Senior Research Scholar in WRM at the Okavango Research Institute of the University of Botswana. He has an MA in Environmental Science from the University of South Africa and a PhD in Natural Resources Management from the University of Botswana. He has published extensively in WRM, livelihoods, household access to capital and adaptation to flood variability and change. His current research interests are water sector reforms and institutional responses to climatic change.

The Botswana site DARMA team visits the unique Tsodilo Hills (UNESCO World Heritage Site) while on a fieldtrip in the Okavango Delta. Credit: Alyne Delaney

Preface

This book is about partnerships and power games in natural resources management (NRM) in the Okavango Delta. Multiple stakeholders and multiple interests are synonymous with the field of NRM, as are contestations over environmental goods and services. As a result, partnerships and power games are inevitable in the interaction and participation of stakeholders in negotiating their stake in NRM. In the Okavango Delta partnerships have been formed to embrace new thinking in NRM as embodied in integrated natural resource management (INRM) and associated concepts; and power games have been played as stakeholder and actors attempt to deal with competing claims over natural resources.

Rural livelihoods in Botswana in general and specifically in the Okavango Delta rely heavily on natural resources (NR), and thus environmental sustainability and access to land and water resources are of paramount importance to rural communities. Many stakeholders are interested in NR: government, for reasons including economic growth; donors and civil society, who lay claims citing conservation; the private sector is interested in the environment as a business commodity; tourists vie for the scenic appeal and therapeutic benefits. The list goes on. Natural resources have become a battlefield and an arena for negotiating access and benefits. In the process partnerships are created and dismantled and power games are played, won or lost. In the Okavango Delta a 'draw' (win–win) is yet to be experienced.

In this book we present the Okavango Delta Management Plan (ODMP) development process as the main arena for negotiating stakes in and access to the Okavango Delta land and water resources and hence as a battlefield for partnerships and power games. We also present other side events in NRM at which mini platforms were created and partnerships were formed and used to negotiate stakes, claim authority and rights over resources and stage power games in a situation of the 'survival of the most resourced'. Thus, it is often those with political, economic and knowledge power who stand a better chance to win.

The book is arranged in two parts: Part I presents the 'process' of NRM, otherwise known as natural resource governance. In this section we present the dynamics of 'who' and 'how' and with 'what consequence'; in some of the chapters the 'how' is unpacked into the 'with and without who'. Part II presents the 'content' of NRM, which deals with the questions of what, when and why. This section deals with the 'stakes' of what resources are contested and why.

In several places, the book concludes that the desirable but elusive 'draw' is postponed by an uneven platform for negotiating stakes; that dispossession – or lack – of economic, political and or knowledge power tilts the playing field often to favour or disfavour specific players. Rural communities are a case in point: often they do not possess – or rather are dispossessed of – power or the source of it. In the Okavango Delta and perhaps all of rural Botswana, land and the resources therein are a potential source of power for communities, yet they have been legislated out of 'ownership' and control. The book does not however suggest that sources of power can be evenly distributed among stake contesters as this is not attainable; instead it proposes that if these be combined through a formula that fosters integration and complementarity, as opposed to fragmentation of stake contesters and stake managers, it would go a long way towards levelling the playing fields.

The main problem in NRM is fragmentation, where sectors put sectoral issues ahead of any others, ignoring the connectedness and relationships within natural systems, as the foreword to this volume contends. This book contributes to the DARMA series in an attempt to defragment NRM in Southern Africa, focusing on the Okavango Delta.

Acknowledgements

First and foremost we thank the funder, EU African Caribbean and Pacific Group of States (ACP Secretariat) for generously funding the Defragmenting African Resource Management (DARMA) Project and the production of this book, and for their contribution to defragmenting natural resource management in Southern Africa in general and the Okavango Delta in particular. We are also grateful to the Institute for Poverty, Land and Agrarian Studies (PLAAS), University of the Western Cape (UWC), in particular Associate Professor Mafaniso Hara for tirelessly and meticulously leading both the DARMA project and the subsequent books project.

Sincere gratitude is extended to the Southern African DARMA (and commons) research team and our European partners who developed the DARMA project proposal under the guidance and leadership of the late Professor Douglas Wilson (Aalborg University, Denmark). We are also extremely grateful to the contributors to this book, who simply made it happen. While we take full responsibility for any lapses in quality we are grateful to all those who dedicated time to review our work and improve the quality of the manuscripts in this book.

We are also grateful to the University of Botswana, for allowing time investment into producing this book; and to our immediate supervisors and colleagues at the Okavango Research Institute and Innovative Fisheries Management for their support. To the people of the Okavango Delta, who took the time to work with us, we are thankful for their willingness to collaborate through sharing their views and knowledge.

Finally we are grateful to our families who rendered all the support they could while they waited patiently for us to complete this book project.

Lapo and Alyne

Dedicated to the people of the Okavango Delta to whom I owe a lot in terms of my professional and personal growth!

Lapo

And to the memory of our colleague and mentor Douglas Clyde Kongshøj Wilson, who is greatly missed.

Alyne and Lapo

INTRODUCTION

1 Successes and Challenges of an Integrated Natural Resource Management Strategy in the Okavango Delta Ramsar Site

Lapologang MAGOLE and Alyne DELANEY

Abstract

Throughout the world, the natural resource management (NRM) process has seen hostile struggles, violence, war, litigation and/or subtle forms of social conflict (especially under open access) as a result of the collapse of a communal management system (Bolaane 2004; Magole and Magole 2009). These have necessitated the development of collaborative management solutions. For the past three decades the agreed vision for new thinking in NRM is: sustainability; sustainable economic development, sustainable livelihoods and equity; and sustainable ecological functions. However, it is still a struggle for development practitioners and conservationists to achieve this trio simultaneously in a win-win-win situation.

Introduction

Borrini-Feyerabend et al. (2004) state that content and process are two key dimensions of NRM. They describe NRM as "responding appropriately to ecological characteristics of a given environment, preserving its integrity and functions, while assuring a flow of benefits from it"(2004: 64), i.e. content – what and when. Secondly, it is "responding to the social characteristics of the same environment, dealing in an effective way with the inevitably conflicting interests and concerns of different social actors"(ibid), i.e. process – who and how. Content therefore is the actual management, defining what is actually done; while process, on the other hand, can be equated with governance – who does what and how. Good natural resources (NR) governance involves striking a balance between content and process to achieve the societal goal of environmental, economic and social sustainability.

Although widely accepted, the concept of sustainability has been viewed with suspicion by both the development (economics) and conservation (environment) camps, each accusing the other of hijacking it. The immediate NRM system solution to operationalise the concept of sustainability is the ecosystems or landscape system, which advocates for a management system which considers existing or traditional land uses alongside new development and conservation of resources. The idea is to create a systemic environment which allows these to exist side-by-side in natural harmony.

The struggle between development and conservation continues and, until recently, the socio-institutional issues of equity and governance were side-lined and ignored resulting in persistent conflicts over natural resources. In Ngamiland, these conflicts include those between traditional livelihoods such as livestock rearing and tourism and human/ wildlife conflicts of crop damage by elephants and livestock predations by lions and other predators. In response, experts and academics have over the past decade been working with other stakeholders in an effort to reconcile conservation and development objectives. In the process however they found that the biggest obstacle is reconciling the multiple stakeholders of NRM, often times with conflicting interests. Thus, as Borrini-Feyerabend et al. (2004) explain, responding to the social characteristics and dealing with inevitably conflicting interests of different social actors is the biggest challenge in NRM, i.e. who and how. This finding has in recent years raised the profile of governance in NRM and resulted in a call for integrated systems of NRM planning; integrated natural resources management (INRM), integrated water resources management (IWRM), adaptive management, co-management and sustainable land management (SLM) to name a few.

Natural resources management planning

A NRM plan is a set of ideals for the sustainable utilization of available resources in a defined area and is hence expected to guide resource users and managers in decision making and prioritisation. However, the plan can be said to fulfil these functions only if it is implemented. According to Conyers and Hills (1986), a plan is more likely to be implemented if it is comprehensive, flexible, adaptable, realistic and perhaps most importantly, acceptable to all stakeholders. Current thinking is that a NRM plan produced through the application of an integrated approach – co-management, ecosystem approach, IWRM – is most likely to achieve high levels of acceptance and implementation and impact. However, experience has shown that this is not guaranteed; conditions that may negatively affect implementation, even under integrated approaches, include the nature and extent of the conflicts over natural resources involved, interpretation of the approach, as well as the socio-political and cultural environment upon which the approach is applied (Bellamy et al. 1999; Swatuk 2005).

There are three distinct groups of players in NRM planning:

a) Community – group(s) of people living within close proximity of the resource(s)
b) Donors – organisation(s) providing financial and material resources for conservation, and sometimes exploitation, of resource(s) to benefit local communities and national economies. These may be:
   - Government
   - Local and/or international donors

  - Local and and international non-governmental organisations (NGOs)
  - Private companies/individuals.

c) Process Facilitator – Organisation(s) providing the human resources and technical expertise to drive the process of resource(s) conservation and exploitation. These may be:
  - Government departments
  - Local and or international donor(s)
  - Local and or international NGOs
  - Private companies/individuals
  - Research /knowledge institutions.

In order to reconcile multiple stakeholders and interests, integrated management systems such as INRM, IWM and the ecosystems approach call for consultation and stakeholder participation and a level platform for negotiation of stake by all stakeholders. The most recent development is to treat NRM and access to natural resources as a human rights issue, evoking concepts of democracy and good governance in natural resources decision-making. For this to be achieved all stakeholders need to have the capacity to participate meaningfully in the form of trained personnel and finance for activities, and a supportive legislative and policy environment. Thus, key capacity indicators are empowerment; do stakeholders have knowledge or access to it (transparency); do they exist in an enabling legislative and policy environment; do stakeholders have resources (financial and human) or access to them to participate in planning and implementation of resource use and control instruments?

This book is an attempt to answer these questions and is a discursive piece about a decade- long planning process aimed at development and conservation of the Okavango Delta.

The study area

The study area for the entire book is the Okavango Delta. The Okavango River and Delta are parts of a river basin system shared between the countries of Angola, Botswana and Namibia (see Figure 1.1). The River Basin system delivers an average of 9.4 cubic km of water into the Kgalagadi Sands at the top northwestern corner of Botswana, turning an otherwise desert area into a wetland of international importance (http://www.ramsar.org/sitelist.pdf; Wolski et.al 2005).

The resources of the Okavango Delta support a diversity of livelihoods ranging from agricultural production to tourism enterprises (Kgathi and Ngwenya 2005; Magole and Thapelo 2005). As the second most visited tourism destination in the country (Magole and Gojamang 2005), the Okavango Delta has the potential to contribute significantly to the national economy. Furthermore, in 1997 the Government of Botswana listed the Delta and its surroundings as a Ramsar site, by so

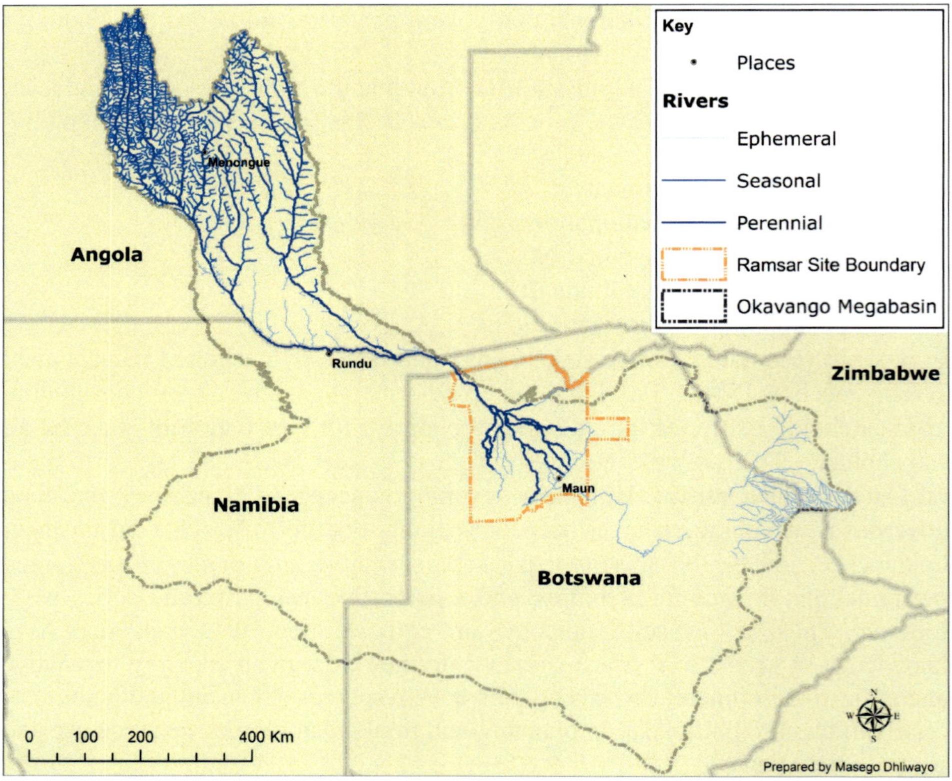

Figure 1.1: Okavango Delta Ramsar site and Okavango RiverBasin. Source: ORI GIS Laboratory

doing giving it international significance and importance beyond its fellow riparian states of Angola and Namibia.

Resource use issues of the Okavango River basin as a whole are made complex by the fact that it is a trans-boundary river system requiring international cooperation and coordination in resource utilisation (Swatuk 2003; Turton, Ashton and Cloete 2003). Owing to its position at the bottom end (downstream) of the Okavango River, system management of the Okavango Delta requires even more prudence and tact than does the rest of the system (Swatuk 2003). In addition, being a Ramsar site means that the international environmental management lens is focused on activities in the Okavango Delta. Locally, resource use is further complicated by the fact that the Okavango River and Delta support a diversity of stakeholders "with different perceptions of land and natural resource utilisation" (Jansen and Madzwamuse 2003: 144–145; Mbaiwa and Darkoh 2005). The two plans referred to in this chapter were developed primarily to address this last point.

Table 1.1: Stakeholders of the Okavango River Basin

#	Stakeholder	Interest
1	Riparian communities	Conservation Livelihoods
2	Resource managers (govt. institutions)	Regulate resource use Conservation Earnings
3	Policy makers (Government of Botswana, OKACOM, SADC)	Regulate resource use Sustainable development Votes
4	Local NGOs	Conservation Community development (livelihoods) Fund raising
5	Business community (tourism and others)	Financial returns Conservation
6	International partners (other Riparian States, Ramsar, etc.)	Sustainable development Conservation Poverty alliviation Millenium Development Goals
7	Tourists/visitors	Leisure/holiday Conservation
8	Researchers/academics	Publications Policy influence Sustainability (including livelihoods)

Natural resource use issues in the Okavango River and Delta

As indicated earlier, the Okavango Delta is a complex hydrological system, the management of which is complicated by the existence of multiple, as well of at times competing, stakeholders (see Table 1.1); hence conflict over the natural resources is inevitable.

There are now more natural resources use conflicts in the Okavango river and delta region than ever before (Mbaiwa and Darkoh 2005; Magole and Demotts, this volume). Farmers complain in particular about increased crop damage by wild animals especially elephants and predation of livestock by wild animals, as well as the shrinking grazing land due to erection of disease control fences.

Local people complain in general about large land allocations for tourism or other commercial purposes that cut them off from their sources of livelihood. There are also fears that the water within the river channels may be polluted by mismanaged sewage and solid waste. There is also conflict among recreational, commercial and subsistence fishers.

Environmental management planning and conservation efforts in the Okavango Delta

The Okavango River Panhandle Management Plan (ORPMP)

The ORPMP is a land and water resources management plan which was initiated by the Tawana Land Board (Ngamiland District land use and planning authority) in an attempt to contain the problems and conflicts about resource use experienced in the Okavango River Panhandle (the river that empties into the Okavango Delta). The plan laid out how the river resources – ranging from the fish to the land – were to be accessed and used. For several reasons many components and provisions of this plan were never implemented.

Since the Land Board was the sole sponsor, the project had a limited financial budget. This translated into both a limited time budget as well as human resources investment. The consultants employed to develop the plan had only six months to complete the task and, for example, the consultancy team could not employ a qualified land-use planner. As a result the plan lacks basic planning characteristics such as goals, objectives, targets, thematic framework, time frames and task allocations as well as the cost implications of implementation. Secondly, the integration claimed by the plan is only limited to the spectrum of natural resources dealt with; it does not deal with the relationships within the resources, nor does it make an effort to integrate resource managing sectors and users. The plan deals with issues of water, wildlife, grazing, arable agriculture, veld products, and fish management and their utilization, albeit as separate entities.

Stakeholder involvement was also limited. The consultant was the sole producer and the Land Board the sole owner of the plan. As stated earlier, the Land Board allocates land but other Departments manage the resources therein, hence the authority of the Land Board over the use and management of NR is limited. Once the plan was produced, other Departments had to implement its provisions. However, this did not happen mainly because the other Departments were not involved in the development of the plan and so were not intimately familiar with it. There is evidence of community consultation meetings. However, from the available records these appear to be superficial and limited in both coverage and scope. There is no evidence of community input into the plan. A review study carried out in early 2007 revealed that 60% of the community leadership had never heard about the ORPMP, and about 80% of those who had heard of the plan could not remember a single of its provisions. None of those interviewed, including resource managers, had a copy of the plan. As a result of limited investment in stakeholder involvement, especially communities, stakeholder buy-in and support for the plan were lacking. A case in point is the provision which required the setting of a 500 meter buffer between users and the river, which was so unpopular that the Land Board could not enforce it. Other provisions were outright ridiculous and were ignored by the Land Board itself; for example it was recommended that the Land

Board stop allocation of fields for arable agriculture but this could never be implemented without widespread consultation as arable farming is a major livelihood source in the Panhandle.

Another salient and important reason why implementation of the plan was limited is that Land Boards are local (District Level) authorities and many resource managing Departments belong to central Government. In Botswana's governance system, central Government is superior to local Government in policy development and indeed all planning and decision making in general; thus the planning system is top-down. Implementation orders that appear to come from the bottom end may not be welcomed at the top end.

The ORPMP was flawed as a plan and as a planning process. The production of this plan may be said to have towed the line of plan development in Botswana; that line is a sectoral system where specialized Departments, authorities and organizations take sole responsibility for otherwise integrated and complex systems. As a result, the plan lacked the all-important stakeholder involvement and support; it ran far short of current NRM thinking which endorses integrated and ecosystems approaches and it lacked investment in the key areas of time, finance and human resources.

Another planning initiative was undertaken in 2003 to produce the Okavango Delta Management Plan (ODMP). Geographically, this initiative involved a bigger area than the ORPMP, covering the entire Ramsar site as shown in Figure 1.1. The Plan also attempted to employ INRM principles to improve the chances of implementation and having traceable impact.

The Okavango Delta Management Plan (ODMP)

Why the ODMP?

The ODMP is an integrated land and water resources management plan for the Okavango Delta Ramsar site. According to the plan proposal document the ODMP aims "to develop a comprehensive, integrated management plan for the conservation and sustainable use of the Okavango Delta and surrounding areas" using the ecosystem approach, which among other things advocates stakeholder involvement in the management of natural resources (Government of Botswana 2002: 12).

The impetus for the ODMP planning effort grew from Botswana listing the Delta as a Ramsar site in 1997. Among other obligations, Botswana, as a contracting party to the Ramsar Convention, has to develop plans to promote *wise use* (Swatuk 2003: 898) of the listed wetland system. Thus the planning initiative was the fulfilment of Article 3.1 of the 1971 Ramsar Convention which states that: "The Contracting Parties shall formulate and implement their planning so as to promote the conservation of the wetlands included in the List, and as far as possible the wise use of wetlands in their territory." At 6,864,000 hectares, the Okavango Delta area is the largest Ramsar site in the world.

Secondly, the planning initiative is within the national ideal of sustainable development as outlined in the National Development Plan (NDP 9), a vision which is carried on into the current NDP 10. Furthermore, the Okavango Delta area has been identified by the country's Vision 2016 as requiring planning in order to promote prudent use of its resources for the benefit of the local communities, the nation and the international community at large. As stated above and articulated in Mbaiwa (this volume), the Okavango Delta is a major tourism destination and hence important for the national objective of economic diversification from a mineral-based economy.

Table 1.2: Analysis of issues affecting the Okavango Delta Region

Issue/Problem	Underlying drivers (causes)	Interests at stake	Issue dynamics
Poverty among riparian communities	– Resource access policies – Development strategy	– Livelihoods – Conservation	– Delayed rural population development – Focus on infrastructure provision – Low literacy rates
Disjointed management of resources	– Lack of integrated management plans/policies (national and basin wide)	– The economic and environmental benefits from basin natural resources	– Sectoral natural resources management system – Contradictory resource policies
Animal and human diseases	– Natural process – Incompatible land uses – Poverty – Risky behaviour	– Biodiversity – Livelihood sources	– Separation of wildlife and livestock
Skewed distribution of benefits from water resources; e.g tourism	– Disjointed management of resource development, – Lack of stakeholder participation in plan and policy formulation and – implementation	– Livelihood sources – Gross Domestic Product (GDP)	– Flawed natural resource development planning processes – Governance structure
Natural resources use conflicts	– Flawed natural resource development planning processes – Lack of communication	– Livelihoods – Profit	– Flawed natural resource development planning processes – Biased resource development and management decisions
Extreme water events (flooding, drought, drying or desiccation of rivers)	– Natural processes	– Livelihoods – Assets	– Land rights

Thirdly, ODMP is a challenge to the existing system for sectoral planning and resource management in Botswana. It is believed that most of the resource use conflicts in the Okavango Delta occur because Government Departments responsible for managing different components of the ecosystem neither communicate with each other or with other stakeholders. Conflicting policies are also a problem due to lack of cooperation and coordination in the management of the Delta resources and NR in the country as a whole. Table 1.2 presents a comprehensive list and analysis of the broader issues affecting the Okavango Delta region. Due to an urgent need for a shift in NR governance and management in order to resolve these, the ODMP emerged as a welcome solution for the parties involved.

Guiding principles for the ODMP

In order to deal with the above problems the ODMP process was crafted to follow the now-popular integrated water resources management (IWRM) approach as recommended by the Ramsar planning guidelines and the IUCN ecosystem approach. Lang (1990: 3) as quoted by Bellamy et al. (1999: 239 and 342) argues that situations demanding integrated approaches "are highly complex, interests and objectives are numerous and in conflict, information and knowledge are quite incomplete, ends and means are ambiguous, control is fragmented and the external environment is in flux". This renders the Okavango Delta area and indeed the whole arena of natural resources management and use in Botswana a good candidate for IWRM.

In its purest and ideal sense IWRM is no doubt the solution to most natural resources management and use problems. The approach is hailed by many (Bellamy et al. 1998; Al Radif 1999; Dungumaro and Madulu 2002; Jonch-Clausen 2004; Van der Zaag 2005) as it has the potential to bring about successful and sustainable water resources management by providing a platform for stakeholder dialogue, cooperation and negotiation for access to scarce resources. According to paragraph 20 of the Ramsar Convention on Wetlands New Guidelines for management planning for Ramsar sites and other wetlands, "the aim of Integrated River Basin Management or Integrated Water Resource Management is to bring together stakeholders at all levels, from politicians to local communities, and to consider water demands for different sectors within the basin" (http://www.ramsar.org/key_guide_mgt_new_e.htm).

Two things seem to make this approach very attractive: the promise of sustainability of the ecosystem and the goods and services that it provides, and the potential for resolving conflict. However it is important to warn that IWRM is no magic or quick solution or generic answer to all natural resources use and management issues (Lovell et al. 2002; Jonch-Clausen 2004). As Dungumaro and Madulu (2002) argue, how much success the approach achieves in a particular water or natural resources management initiative depends on numerous factors. These include the nature and extent of the NR conflicts involved, interpretation of the approach, and the socio-political and cultural environment upon which the approach is applied (Bellamy et al. 1999; Swatuk 2005).

Basic planning requirements of financial and human resources are very important, but these are almost always provided. They were certainly provided for the production of the ODMP. However there is a tendency to ignore the role played by political will and institutional arrangements. IWRM requires institutional arrangements which provide equal opportunity for all to negotiate their stake. By implication IWRM is unlikely to happen if institutional arrangements remain empowering to some and disempowering to others. The question is, does Botswana have an institutional framework for natural resources planning which provide an enabling environment for an integrated approach such as that attempted by the ODMP?

Swatuk (2005) rightly warns that the prerequisites of integrated approaches are much tougher than those of the traditional (sectoral) system of management, as these approaches integrate, or rather blend, everything and everybody. The most difficult prerequisite is that the last be put first (Chambers 1983), that voiceless rural communities be given a platform to speak and be listened to by resource managers who may not have the culture, attitude or will to do so. More than anything, integrated approaches promise to tackle the real natural resources problems of power, equity and access (Van der Zaag 2005). Herein lies the problem, according to Swatuk, who argues that "this is a profoundly political act which challenges the very basis of power in many societies" (2005: 878). Levelling the playground like this is likely to be met with resistance by those who hold power.

The ODMP planning process

As a working development strategy, the ODMP was initially divided into 12 components, each representing a prominent resource use sector such as water or river channels, fish, land, plants and range, wild animals, livestock, tourism etc. It is important to mention the existence of a less conventional planning sector of stakeholder participation, research strategy and data management components. In most planning processes these are considered cross-cutting issues and hence inherent and to be addressed within the other sectors/components. The end result is that sector specific issues take precedence in terms of time and allocation of other resources and these so-called "cross cutting issues" (stakeholder participation, research strategy and data management) are marginalised. However in the ODMP process these formed a substantive sector, complete with a budget, and allocated to an equally unconventional but appropriate planning partner, the University of Botswana's Harry Oppenheimer Okavango Research Centre (HOORC), now the Okavango Research Institute (ORI) (Magole and Demotts, this volume). This is to be applauded as planning processes often miss out on the contribution of academics.

The sectors were each tasked with situation analysis within their sector to answer questions including: What is the state of the resource? Who is using the resource? Who is managing the resource? Who else has interest in the resource? Are there any conflicts in the use of the resource? To operationalize these points, integration sectors were organized into task forces around each planning component. Task forces had members from all other natural resources sectors, professionals and user groups of the resources within the Delta. Thus, multidisciplinary teams were formed in accordance with the requirements of IWRM (Al Radif 1999). This allowed for a rare opportunity for the sectors to communicate their views and plans, learn from each other and to negotiate their stake in other sectors. The tourism sector, for example, was able to negotiate its requirements for land and other resources with the Land Board and wildlife sector. Communities were also brought into the process through the inclusion of community leaders, resource use group representatives, as well as community focal persons who became members of task forces and acted as the intermediaries between the ODMP technical planning team and the communities. Workshops were organized to provide a forum for the community representatives to contribute their perspective on the situation of the resources, to suggest solutions and to appraise solutions suggested by the technical planning team.

The strategy for managing and or supervising the plan production also contained elements that are unconventional within the country's planning system. The planning process was led and coordinated by the Department of Environmental Affairs) under the Ministry of Wildlife, Environment and Tourism. The coordinators (DEA) reported to a Steering Committee of Permanent Secretaries and Directors of all Departments involved. At the District Level the coordinators and the planning sectors reported to the Okavango Delta Wetland Management Committee of all District level stakeholders and task forces of experts as described earlier. Last but not least the coordinators and the sectors had to report every milestone achievement to the communities in the village *Kgotla* (traditional meeting place and the seat of the *Kgosi* or Chief). Altogether this strategy gave the ODMP a stakeholder involvement profile second to none in the country.

Did the integrated natural resource management approach aid implementation of the plan in the Okavango Delta region?

The integrated nature of the ODMP was supported by the prevailing culture of integration at District level. The ODMP Secretariat and planning team were to the plan production process what the office of the District Commissioner (DC) and the District Development Committee (DDC) are to the District Development Plan (DDP) production process. The Secretariat promoted horizontal lines of communication. However, the Secretariat faced some challenges even before implementation during the otherwise glossy planning process.

There were problems with 'assigned' officers showing more loyalty to their respective Departments than to the ODMP process. The ODMP process began to suffer the consequences of adopting an integrated approach within a sectoral institutional setting. The fact that officers were 'assigned' and not 'seconded' to the ODMP posed a planning constraint. The ODMP Secretariat did not control the officers' time and budget. In fact participating Departments had the liberty to make spending and priority decisions 'almost' independent of the Secretariat. The Secretariat could only query or question extreme deviation from the project objectives. A case in point was the water component; the Department involved chose to use most of its budget on developing a hydrological model at the expense of the stakeholder issue of concern with blocked river channels which were no longer delivering water to their agricultural areas. Addressing local stakeholders' issues was key to ODMP but not so for the Water Affairs, at least not at the time. Negotiations for a better approach proved futile because at the end of the day the Secretariat could not force the preferred management approach on a department with a legal mandate to make decisions on water management.

With much effort and persuasion from the Secretariat an integrated plan was produced. However, as happens with the DDPs, the sectoral system resurfaced at the stage of implementation. Government Departments have the autonomy to make decisions and priorities for implementation. It turned out that many of the sectoral priorities or urgent issues were not the same as those identified under the ODMP. This is a limitation caused by the broader national planning system from which no planning effort is able to deviate. In addition, the emphasis was on national scale planning rather than at a local. Thus while the ODMP managed to raise international support for the planning phase, resources for implementation depended on Government spending which does not sponsor region/district specific planning effort traditionally.

Even the ODMP 'Secretariat' may be said to have reverted to the sectoral approach. The project Secretariat has disbanded and has been replaced by the less resourced and less powerful Department of Environmental Affairs). The Department simply added to its many duties as an environmental coordinating agency the duty to 'encourage' other Departments to implement ODMP provisions. There is doubt that this will bear fruit significantly. Clearly a properly functioning and well-resourced agency is required for implementation if the ODMP is to achieve its objectives and if the ideals of an integrated approach are to be achieved.

Conclusion

The Okavango Delta management 'team' of experts acknowledge that implementation of the integrated ODMP has been challenged by the inherently fragmented institutional system within which it was conceived. Admittedly, NRM is a political process which is charged with conflict over control and access to scarce resources. However it is suggested that such a fragile and highly contested system as the Okavango Delta requires a planning authority with power and resources to plan for and manage it as an ecosystem. Hence socio-political issues of power, equity and social justice need to be resolved if such innovative initiatives as the ecosystem approach, co-management and IWRM are to stand a chance of success.

The rest of this book leans heavily towards evaluation of the ODMP as an indirect way to illuminate the challenges of implementing integrated natural resources management approaches; perhaps this is premature for most governance systems in Southern Africa. However, compiling this volume within the context of the DARMA project brought together natural resources experts to appreciate the rather similar issues and challenges to natural resources management facing the environment industry as a whole and underlined the need to act together to defragment natural resources in Southern Africa. The book is organised in two complementary parts as follows:

Part I: Governance Issues

Chapter 2 presents a critique of stakeholder participation in the ODMP. One of the hallmarks of the ODMP was its drive for stakeholder participation. This chapter provides a critical analysis of the platform for participation created and questions whether all stakeholders had the same opportunity to participate and negotiate their stake. It uses a post-mortem stakeholder analysis to determine which stakeholders succeeded and failed to negotiate their stake.

Chapter 3 discusses the role of the University as a supposedly neutral stakeholder during the ODMP planning process. It presents the role of the University of Botswana Okavango Research Institute, then the Harry Oppenheimer Okavango Research Centre, in the planning process and how it was perceived by other stakeholders, especially communities. It analyses critically the role of the University as an empowering partner with communities which are often the weaker partner.

Chapter 4 deals with stakeholder participation in NRM in the Okavango Delta, with particular focus on the internal dynamics within communities. It uses the 'dash board' tool to measure the accountability of a community natural resource committee to its own members (the community). The analogy of a dash board uses warning lights (green, amber and red) to warn the community of the danger of lapses in accountability, transparency, financial (mis)management and democracy or good governance. The chapter provides the results of applying this tool in Tubu Village.

Chapter 5 focuses on community-ecotourism and presents a case study to explore whether communities get a 'good' or 'raw' deal in tourism partnerships. The chapter presents a case study of a community venture and its partnership dynamics. The case provides an example of a management gap within NRM regulatory and practice framework, where government (or other players such as civil society) fail to protect the rights of communities involved in joint ventures with more powerful partners.

Part II: Natural Resources Management and Use Issues

Chapter 6 presents a case study of the implementation of the Management Oriented Monitoring System (MOMS) as a conflict resolution tool in NRM for an agro-pastoral community living in the Okavango Delta. Earlier chapters have referred to communities as the weaker partner within a spectrum of stakeholders. One of the weaknesses identified is a lack of information. This chapter presents the result of an initiative to empower the community of Tubu with a natural resources monitoring system which enabled them to collect and analyse data on different aspects of their resource use in order to make informed decisions and negotiate with other stakeholders. The chapter shows how the tool worked as a means of conflict resolution in natural resources management.

Chapter 7 presents an analysis of the drivers of tourism as a resource use and development activity in the Okavango Delta. The analysis illustrates Government vision and the policy environment and examines if these provide appropriate space for communities to participate and benefit significantly from the booming tourism industry. It concludes by looking at whether the ODMP makes a difference for communities in the Okavango Delta.

Chapter 8 presents an assessment of the agriculture water resources use in the country in general, and in the Okavango Delta Region in particular. The Okavango Delta area is perhaps the most water-endowed area in Botswana. This chapter looks at water resources management efforts, if any, for agricultural development. The chapter places the Okavango Delta within the national irrigation situation and asks whether sufficient use is being made of the area's comparative advantage to produce food for the region and nation.

Chapter 9 presents a broad assessment of small-scale fisheries (SSF) management in floodplain systems using the case of the Okavango Delta. It asks the questions: What needs to be managed? How should it be managed? The chapter describes the evolution of fisheries management in SSF, using the Okavango Delta as a case study. It critically analyses fisheries management from a (elitist) classical perspective which is deeply rooted in resource 'preservation' for the benefit of a select few and has been the dominant system for many developing countries including Botswana. The chapter compares and contrasts the classical fisheries management with indigenous traditional management whose basic tenets are sustainable utilisation and conservation (based on egalitarian principles).

The strengths and weaknesses of each approach will be explored, and a hybrid management regime proposed based on these two approaches. The chapter concludes with a proposal for a paradigmatic shift in SSF management based on ('inclusive') holistic approaches versus ('exclusive') top-down approaches.

Chapter 10 concludes the book by zooming in on four issues that appear to cut across all eight substantive chapters: Governance, stakeholder participation, local knowledge and co-learning and the participation of interested 'outsiders'. It concludes that while local resource users are disempowered by being left out of NRM structures, there are structures at the local (District) level which can be supported, with the right policy and legislative environment and resources, to take control of NR governance in the Okavango Delta. It proposes that integrated natural resources management (INRM) and hence *sustainability* will only be achieved if resource users are empowered to participate in these local structures.

References

Adil, Al. Radif (1999). "Integrated water resource management (IWRM): An approach to face the challenges of the next century and to avert future crises." *Desalination* 124(1999) 145–153.

Bellamy, A. J. et al. (1999). "Policy Review: Evaluating Integrated Resource Management" *Society and Natural Resource*, Taylor and Francis; Brisbane.

Bolaane, M. (2004). The Impact of Game Reserve Policy on the River BaSarwa / Bushmen of Botswana. *Social Policy and Administration*, Vol. 38, No. 4, 399–417.

Borrini-Feyerabend, G., Pimbert, M., Taghi Farvar, M., Kothari, A. and Renard, Y. (2004). *Sharing power, learningby doing in co-managementof natural resources throughout the world*, IIED and IUCN/CEESP/CMWG, Cenesta, Tehran.

Chambers, R. (1983). *Rural Development: Putting the last first*. John Wiley and Sons, New York.

Dungumaro, W. E. and Madulu, F. N. (2002). "Public Participation in Integrated Water Resource Management The case of Tanzania". Presented at the 3rd Waternet Warfsa Symposium, Dar es Salaam, 30–31 October 2002.

Jonch-Clausen, T. (2004). "Integrated Water Resource Management (IWRM) and Water Efficiency Plans by 2005", Global Water Partnership, Sweden.

Kgathi, D.L. and Ngwenya, B.N. (2005). "Community Based Natural Resource Management and social sustainability in Ngamiland." *Botswana Notes and Records,* Vol. 37, 61–79. Botswana Society, Gaborone.

Lovell, C. (2002) "The question of scale in integrated natural resource management. Center for Ecology and Hydrology." Resilience Alliance.

Magole, L. (2003). "A tragedy of the commoners: The evolution of communal rangeland management in Kgalagadi, Botswana". Ph.D. Thesis. University of East Anglia, Norwich, UK.

Magole, I. L. and Gojamang, O. (2005). "The dynamics of tourism visitation to National Parks and game reserves in Botswana." *Botswana Notes and Records,* Vol. 37, 80–96. Botswana Society, Gaborone.

Magole, L. I. and Magole, L. (2007). Is ecotourism the appropriate strategy for the socio-economic development of Basarwa (Khoisan); the case of Gudigwa village, Okavango Delta, Botswana, in Ahmed, A. (2007) *Managing Knowledge, Technology and Development in the Era of Information Revolution*, World Association of Sustainable Development, University of Sussex, Brighton. (pp. 261–269).

Magole, L. and Thapelo, K. (2005). "The impact of extreme flooding of the Okavango River on the livelihood of the molapo farming community of Tubu Village, Ngamiland sub-District." *Botswana Notes and Records,* Vol. 37, 125–137. Botswana Society, Gaborone.

Mbaiwa, J.E. and Darhoh M.B.K. (2005). "Sustainable development and natural resource competition and conflicts in the Okavango Delta." *Botswana Notes and Records,* Vol. 37, 40–60. Botswana Society, .

Ramsar, Iran. (1971). The Ramsar Convention Manual. "A Guide to the Convention on Wetlands". Gland, Switzerland.

Swatuk, L. (2005). "Political challenges to implementing IWRM in Southern Africa." *Physics and Chemistry of the Earth,* 30, 872–880.

Swatuk, L. (2003). "State interests and multilateral cooperation: Thinking strategically about achieving 'wise use' of the Okavango Delta system." *Physics and Chemistry of the Earth,* 28, 897–905.

Turton, A., Ashton, P. and Cloete, E. (2003). (eds) *Transboundary rivers, sovereignty and development: Hydropolitical drivers in the Okavango RiverBasin.*African Water Issues Research Unit (AWIRU) and Green Cross International (GCI), Geneva.

Van der Zaag, P. (2005). "Integrated water Resources management: Relevant concept or irrelevant buzzword? A capacity building and research agenda for Southern Africa." *Physics and Chemistry of the Earth,* 30, 867–871.

Wolski, P. et al. (2005). "Aspects of seasonal dynamics of flooding in the Okavango Delta" *Botswana Notes and Records,* Vol. 37, 179–195. Botswana Society, Gaborone.

PART I – GOVERNANCE ISSUES

2 Who Got Heard During the Okavango Delta Management Plan (ODMP) Process? A Critique of Stakeholder Participation in the ODMP

Lapologang MAGOLE and Phemo Karen KGOMOTSO

Abstract

The Okavango Delta Management Plan (ODMP) was first and foremost initiated in conformity with the Ramsar Convention requirement to produce a management plan for listed wetlands. It was, however, also a response to the national ideals and vision for sustainable development. As a natural resource (NR) planning process, the ODMP is also one of the few attempts to operationalise integrated water resources management (IWRM) in Southern Africa. The planning process provided a rare platform for dialogue between nearly all stakeholders of the Okavango Delta and River basin. Stakes were laid down and negotiated.

In this paper we present the results of a study which aims to analyse the resultant policy documents and decisions on the management of the land and water resources of the Okavango Delta. We ask, what changed, if anything? How and why did it change? What are the implications of the changes? The study involved participatory observation of the planning and fledgling implementation processes, a stakeholder analysis, open-ended interviews with several stakeholders, and document review and analysis of the ODMP component plans. We use the results to analyse the platform for negotiation and measure the level of success for different stakeholders to negotiate their stake. We probe and attempt to answer the question of whether all those who needed to be heard were heard? We consider whether power, be it political or economic, played a role in who was heard in the end. We find that the process had benefits for everyone involved but in the end the Government of Botswana and the international stakeholders who possess both the political and economic power were heard.

Introduction

Development of an integrated management plan for managing the Okavango Delta resources is a central component of listing the Delta as a Wetland of International Importance, as per the Ramsar Convention on Wetlands (1971). The development of the Delta's management plan demonstrates the commitment to sustainably manage its resources and improve the socio-economic wellbeing of its people (ODMP Wetlands Conference Report, 2008). Botswana's Government policy also recognises the socio-economic value of the Okavango Delta as an important ecological resource, rich in biodiversity resources that are central to the livelihoods of those dependent on it, as well as its role in Botswana's lucrative tourism industry, which contributes around 5% to the GDP (Ministry of Finance and Development Planning, National Development Plan 10). The resource therefore requires conservation, sustainable utilisation and management. The Government of Botswana's interest in developing the ODMP project was therefore also a response to national goals on sustainable development as expressed in the National Development Plan 9 (NDP 9), and the nation's Vision 2016 (Magole *et al.* 2008).

The process of developing a management plan, as per the Ramsar Guidelines, requires a comprehensive stakeholder involvement process that facilitates the incorporation of the views of all with a stake in the resource, particularly the primary resource users. As emphasised by the Ramsar Convention Secretariat (2007), the involvement and understanding of local communitiesand indigenous peoples, as the custodians and managers of the wetland, in its management is of particular importance as the management planning process should not be seen as having been imposed from outside upon those who depend on the wetland for their livelihoods. This is also in line with the principles of the Integrated Water Resources Management (IWRM) approach, which advocates for stakeholder participation in the decision-making process on the use of resources (Twomlow *et al.* 2008).

The aim of this paper is to study the outcomes of the ODMP planning process and analyse the resultant policy products and decisions for management of the land and water resources of the Okavango Delta. We ask, what changed, if anything? How and why did it change? What are the implications of the changes? These questions contribute towards answering the main question posed in this chapter: who actually got heard during the ODMP process? Knowing the answer to this question shows who, ultimately, controls the Delta and/or benefits the most from it. We believe that it follows that whoever holds power had control of the process and changes it brought about; they are likely also to reap the most direct or indirect benefits, and that power puts them in an advantaged position to either maintain the status quo or negotiate for more control of the resource.

In the following sections, the research methodology is presented, followed by presentation of the theoretical framework upon which the arguments are based. This is followed by the analysis and results section and then the discussion and conclusions drawn from the study, highlighting the significance and implications of the study results.

Methodology and study area

Study area

The study is based in the Okavango Delta Ramsar site area, as shown in Magole and Delaney (this volume). The Okavango Delta is an inland delta that is home to rich biological diversity of 'global significance' and is the focal point of Botswana's (eco) tourism industry (Mbaiwa; Magole and Magole, this volume). The Delta is also home to over 124,000 residents who utilise the natural resources to varying degrees for their livelihoods. As a result of its diversity, the Okavango has in the last decade attracted many land users and other stakeholders whose divergent interests lead to conflict as they compete for access and control (Darkoh and Mbaiwa 2009). A number of communities, for example, have clashed with tourism operators who have been allocated land on which these communities hunted, collected veld products and grazed their livestock. By closely studying the process of the ODMP stakeholder consultation exercise and its outcomes, we aimed to measure the level of success for different stakeholders in negotiating their stake in the resource. A variety of methods were adopted for arriving at the conclusions for 'who got heard during the ODMP process?'

Methods of data collection

This work draws on varying degrees of involvement of both the authors in the ODMP planning process at different times during the process. The specific methods used for data collection and analyses were:-

a) *Participant observation and multi-sited fieldwork:* Since the inception of the ODMP planning process, both authors have observed (one as an internal participant and other external) and studied the process up to its conclusion;
b) *Stakeholder Analysis:* A number of stakeholder analysis tools used previously by one of the authors (see Magole 2003 and 2008) were modified to reflect the issues under discussion in this chapter. This stakeholder matrix lists the stakeholders identified by the ODMP planning process; their interest in the resource; their relative influence over policy; some of the general/specific concerns raised about the use of the resource during the ODMP; and how their concerns are addressed by the ODMP process products;
c) *Key informant interviews*: These were conducted between July 2008 and April 2009, and involved formal interviews; informal discussions or structured conversations; focus group discussions with resource user groups (for example, fishers, subsistence farmers, basket weavers), field visits with resource users (for example, grass collection trips, visits to farmers fields) and structured interviews with resource managers at the local level (government extension officer based in the Okavango Delta), those based in the administrative district of Maun, and the policy makers based in Gaborone, the capital of Botswana. Some interviews were conducted with international stakeholders working on either the Okavango Delta or the broader Okavango River Basin. These interviews addressed different issues of resource use and management at the different levels of the resource;
d) *Document content analysis:* This aspect of the study entailed closely analysing the contents of the ODMP component plans and related document-

ation that informed the planning process, or resulted from it. The main documents studied included the 2005 ODMP Inception Report, the 2006 ODMP Communication Strategy, and the 2008 Final Okavango Delta Management Plan.

Literature review and conceptual framework

The political ecology of resources use in the Okavango Delta

The argument made in this chapter is based on a political ecological thesis that natural resources are contested and access to these is shaped by power relationships between different groups in any one setting. Robbins (2004: 12) explains political ecology as "empirical, research-based explorations to explain linkages in the condition and change of social/environmental systems, with explicit consideration of relations of power". Central to the political ecology debate is the role of social relations and political power in the issues of access and control of natural resources as well as the unequal distribution of costs and benefits from environmental exploitation (Bryant and Bailey 1997; Forsyth 2003; Robbins 2004; Peet and Watts 2004). According to Watts (2000), political ecology aims to understand the complex relations between nature and society through a careful analysis of the forms of access and control over resources and their implications for environmental health and sustainable livelihoods. Watt's definition aims to explain environmental conflict in terms of struggles over 'knowledge' power and practice and 'politics', justice and governance (2000).

Robbins (2004) outlines four theses that political ecology seeks to address: degradation and marginalisation; environmental conflict; conservation and control; and environmental identity (see Table 2.1).

For the purposes of this study, and perhaps most relevant to the case under study, the 'environmental conflict' and the 'conservation and control' theses are probably the most relevant for illuminating socio-ecological issues in the Okavango Delta. As the 'environmental conflict' thesis argues, increasing scarcities produced through resource enclosure or appropriation by state authorities, private firms or social elites accelerate conflict between groups (gender, class or ethnic). Similarly, environmental problems become 'socialised' when local groups (gender, class or ethnic) secure control of collective resource at the expense of others by leveraging management interventions of development authorities, state agents or private firms. So too, existing and long-term conflicts within and between communities are 'ecologised' by changes in conservation or resource development policy (Robbins 2004).

The 'conservation and control' thesis argues that the control of resources and landscapes has been wrested from local producer groups (class, gender or ethnic) through the implementation of efforts to preserve 'sustainability', 'community' or 'nature'. In the process, local systems of livelihood, production, and socio-

Table 2.1: Four theses of political ecology and what they attempt to explain

Thesis	What is explained	Relevance
Degradation and marginalisation	Environmental change: why and how?	Land degradation, long blamed on marginal people, is situated in a larger political and economic context.
Environmental conflict	Environmental access: who and why?	Environmental conflicts are shown to be part of larger, gendered, class and race struggles and vice versa
Conservation and control	Conservation failures and political/economic exclusion: why and how?	Usually viewed as benign, efforts at environmental conservation are shown to have pernicious effects, and sometimes fail as a result
Environmental identity and social movement	Social upheaval: who, where, and how?	Political and social struggles are shown to be linked to basic issues of livelihood and environmental protection

(Adapted from Robbins, 2004:14)

political organisation have been disabled by official and global interests seeking to preserve the 'environment' (Robbins 2004).

The newer political ecology draws from concern within post-structuralism with knowledge, power and discourse: what individuals and groups (and de facto communities) know and practice with respect to their local environments (Peet and Watts 2004). Political ecology therefore "opens the possibility for a serious discussion of how nature and environmental problems are represented and the discursive formations which shape policy and practice" (Peet and Watts 1996; 2004: 10). Policy can be understood as discourse (Hajer 1998; Keeley and Scoones 2003). "Discourse and the ability to shape discursive terms deeply influence entire frameworks of resource access. By speaking of a 'global commons,' for example, "powerful international NGOs and other actors create universalising categories and naturalise their interventions around the world in the name of environmental protection. Resource access is in this way shaped by the power to produce categories of knowledge (Foucault 1978, cited in Ribot and Peluso 2001: 169). Bryant and Bailey (1997) also highlight the interests and actions of place and non-placed based actors in environmental conflict as key in shaping institutionalised access and control over resources and the associated costs and benefits. For instance, even though institutions like the IUCN and the Ramsar Bureau are located in the global North, their promotion of particular management approaches and interventions through funding or technical support (usually through technical advisory services), ensures that their interests are represented in local level processes. Sometimes the process of implementing/institutionalising these approaches (such as the Ecosystem Approach) is seen by organisations or institutions of water management 'as a largely technical process,' with no recognition that it is a quintessentially political process (Wester *et al.* 2003).

Political ecology is therefore concerned with the "political role of different types of actors in human-environmental interaction" (Bryant and Bailey 1997: 24). Complementary to the political ecology debate is the 'Theory of Access', presented by Jesse Ribot and Nancy Lee Peluso (2003). Ribot and Peluso (2003) define access as "the ability to derive benefits from things" and they state that access here is more akin to "a bundle of powers", locating these 'powers' within the social and political-economic contexts that shape people's abilities to benefit from resources (Ribot and Peluso 20013: 173). Ribot and Peluso's concept of access "aims to facilitate grounded analysis of who benefits from things and through what processes they are able to do so" (2003: 173). As the authors argue, "people and institutions are positioned differently in relation to resources at various historical moments and geographical scales" (2003: 154). Access analysis also helps us understand *why* some people or institutions benefit from resources, *whether or not* they have rights to them.

Stakeholder participation in the ODMP

The ODMP stakeholder consultation process was the first of its kind to be conducted during a planning process in Botswana. While the ODMP project demonstrated commitment to the consultation process, it should also be noted that this process was designed and budgeted for as a core component of a project, almost entirely funded by non-Botswana Government funds, by the German Development Service (DED). The Ramsar Bureau positively encourages stakeholder involvement in the development and delivery of management and decision-making for protected wetlands. As stated in the ODMP Inception Report, "stakeholder engagement has become a fundamental principle of management planning for all Ramsar sites and as a result, stakeholder involvement is also one of the 13 main ODMP project principles" (2005: 30). The ODMP consultation process was also a component of a larger planning process, through which other views were solicited and considered through various reports, workshops and meetings held at various levels during the planning process.

The ODMP Inception Report defines primary stakeholders as those "groups or individuals directly dependent on, and who derive immediate benefit from the natural or other (e.g. cultural) resources of the Delta, whether at a subsistence or small-scale commercial level" (2005: 26). It is recognised that as primary users of the resource, their wellbeing depends on their ability to continue deriving benefits from the resource and therefore the need to maintain the good state of the resource. These primary resource users, and hence primary stakeholders, are the 120,000 residents of the Okavango Delta. Other riparian communities in Angola and Namibia were considered to be secondary stakeholders, not because of their dependence on the Delta but their contribution to the state of the resource. All other stakeholders, such as government institutions, donors, tourists and so forth, were classified as tertiary stakeholders because their dependence on the Delta is minimal.

The ODMP planning process adopted an issue-driven approach of identifying issues of importance to the stakeholders to facilitate deliberation on joint management proposals from users, managers and policy makers. For this purpose, approximately 18% of the Ngamiland population attended the ODMP *Kgotla* meetings held between January and May 2005, with decreasing attendance in major villages such as Maun and Gumare (Kgomotso and Swatuk 2006). Subsequent workshops were held in villages such as Shakawe, Gumare and Maun for interaction with community representatives (Magole *et al.* 2008). At these workshops, most villages in the entire Delta were represented, usually by more than one individual. While acknowledging the limitations of the traditional *Kgotla* system in capturing 'all' views, it should be noted that the ODMP stakeholder consultation exercise was the first of its kind in attempting to conduct a comprehensive consultation of especially primary (rural) stakeholders and actually take their views on board. The consultation process also revealed flaws with the current planning system of Government which is, for the most part, highly centralised.

Stakeholder participation in development planning has often been dismissed as a necessary but costly exercise, resulting in very little meaningful participation in practice. However, stakeholder participation in the ODMP has been praised as 'close to meaningful' and actually capturing the views of those consulted. As noted by an observer and participant in the ODMP planning process, "the ODMP itself was good as it was quite participatory and the issues taken up were those raised by communities as important" (Former ODMP Chief Technical Advisor). Joint planning and consultation among district-based Government institutions inculcated an appreciation of the utility of the integrated planning approach. The different resource management sectors at a District level continue to collaborate as they recognised similarities in each other's work and the need to communicate (Ngamiland District Environmental Coordinator, Department of Environmental Affairs

Villager speaking about animal agriculture conflicts, DARMA workshop. Credit: Alyne Delaney

Results and analysis

Table 2.2 provides an analysis matrix showing the main stakeholder groups identified by the ODMP (2008), the issues raised during the stakeholder consultation process and how these issues have been or are being addressed through the ODMP and subsequent implementation plans. The column on 'current status/results' reflects the views and opinions of the authors about changes, if any, in the situation on the ground as a result of the ODMP.

Based on our observation of the past and current socio-political processes in the Okavango Delta and the broader basin, we have identified four main stakeholders who are relevant to this discussion: local communities (classified as primary in Table 2.2); local tourism industry and national Government institutions (both classified as secondary stakeholders); and the international community (classified as tertiary stakeholders). These stakeholders have varying interests and derive different benefits from the use and management of the Okavango Delta. They also have varying influence on the decisions taken about the Okavango Delta. The outcomes are therefore varying for these different groups, corresponding with their power to influence decision-making about the resource. In the section that follows, we present an analysis of the extent to which each of the stakeholders 'got heard' or was represented in the ODMP planning process and its products.

Table 2.2: Analysis matrix for the ODMP stakeholders, interests, issues and achievements

Primary Stakeholders	Interests	Issues raised during the ODMP consultations	How the issues are addressed through the ODMP or national policy post-ODMP	Current status (Results)
Fishers	Access to fishing for livelihoods *(small-scale commercial sectorand householdsubsistence*)	Conflicts between different fishing sectors (mainly recreational vs. small-scale commercial) over fishing grounds Need to control the sizes of boats used by recreational fishing sector as they disturb spawning fish and bird's nests	– Introduction of Fish Protection Regulations (May 2008) -Introduction of fishing licenses and permits, and a closed fishing season – Revival of the Okavango Fishing Association for different sectors to discuss issues of concern (stakeholder platform)	– Issue of conflict over fishing grounds not addressed by policy, no fishing grounds zoning – Boat sizes only partially regulated (*Fisheries officers not able to enforce the regulationdue to political influence)* – Commercial fishermen still denied access to traditional fishing grounds
Subsistence farmers	Access to agricultural land for food production, and pasture for livestock	Human-wildlife conflicts (elephants raiding crop fields and predators killing livestock both at the river front and in the dry-land grazing areas)	-DWNP pepper project only implemented in some parts of the human/elephant conflict area (Mohembo-Gudigwa stretch) due to limited resources – Government aid announced in late 2008 (just before 2009 elections) to, among other things, fence farmers' fields for free (program never implemented)	– Human-wildlife conflict remains high as very few farmers have adopted/know about the chilli pepper strategy – More farmers exiting arable agriculture and falling into poverty (e.g. abandoning of fields after elephant raids)

Community-based organisations (CBOs)	Access to the tourism industry for local people.	– Citizen participation and benefits from tourism currently non-existent or insignificant. – Inequitable benefits from community-private sector partnerships	-Community Based Natural Resource Management policy introduced in late 2007 to promote community participation in tourism – Decision by central Government to direct 65% of proceeds made by CBOs to a Government controlled Environment Fund	– Policy not yet implemented – CBOs still largely disempowered to negotiate for an equitable stake with the private sector players in the tourism industry as there is no protection from policy Threat of more losses from 65% levy
Secondary stakeholders	Interests	Issues raised during the ODMP consultations	How the issues are addressed through the ODMP or post-ODMP initiatives	Current status (Results)
Tourism business/ tour operators	Access to the Delta resource for business (profit) and conservation	– Conflicts in the fishing sector (lodge/ camp owners complain about small-scale commercial fishermen fishing too close to their properties and overfishing)	Introduction of fishing licenses prohibiting small-scale fishermen to access traditional fishing grounds without permission.	– The tourism sector has managed to influence policy changes within the fishing sector that maintains their access to the resource – Introduction of fishing licenses – Introduction of closed fishing season
Policy-makers (politicians and top civil servants)	Regulation of use of resources Personal access to resources	The need for better understanding of the Delta system and impacts of changes to better manage use of resources	Collaboration with the University of Botswana (UB) Sector policy instruments produced, e.g. land use plan and tourism policy in place	Okavango Delta Information System (ODIS) set up at UB giving access to socio-economic, ecological and hydrological information Some policy instruments being implemented. More government control in the tourism industry.

Government institutions	– Resources management – Promotion of tourism sector	– Need to carry-out institutional mandate – Implement government policy and carry out directives	Institutional strengthening; DEA has set up office in Maun to coordinate the implementation of the ODMP and other environmental management instruments	Several regulatory instruments put in place. Fisheries regulation in place; Tourism development key component in national and district development
Tertiary stakeholders	Interests	Issues raised during the ODMP consultations	How the issues are addressed through the ODMP or National Policy post-ODMP	Current status (Results)
Government institutions from upstream countries	Technical collaboration, co-management, benefit sharing	– Need for co-operation and communication on management of the basin – Need to share benefits from use of the basin resources	ODMP products used in the development of the basin TDA	Technical collaboration continues through specific programs/projects though generally weak and could be improved
International donor agencies (USAID, SIDA , DANIDA, DED)	To promote specific management approaches	Need for primary stakeholder participation and sustainable natural resource management and use	Donor-initiated interventions supported; e.g. DANIDA sponsored ODMP summary booklet (in Setswana and English) to improve communication and implementation	International approaches to use and management of the Delta resources in use through ongoing projects
Tourists	Access to recreational and aesthetic products of the Delta	Need to maintain the pristine and wilderness state of the delta.	– Conservation key theme of the ODMP – Land use promotes tourism.	Tourism (and conservation) remains the main focus of resource utilisation in the Delta
International NGOs (Ramsar Bureau, IUCN)	Conservation ideology	The need to adopt the ecosystem approach and IWRM as planning and management approaches in the Okavango Delta	Ecosystem approach and Ramsar Planning Guidelines used as planning tools during the ODMP	Conservation (and tourism) continues to be the main focus of Okavango Delta resource management

Matrix format adopted and adapted from Magole (2008)

Though the ODMP stakeholder consultation process mainly focused on the Okavango Delta-based stakeholders, especially so-called 'primary stakeholders' (i.e., communities), it is clear from the analysis of the discourse of the ODMP that the views and interests of other 'non-place based' stakeholders were highly influential in the thinking and planning process. This is so, despite the arguments that the ODMP largely reflects the views of primary stakeholders and, to an extent, those of Ngamiland District-based Government institutions. In the section that follows we present stakeholder group-specific analysis of the outcomes of the ODMP process.

Local communities

As noted above, this group of stakeholders depends directly on the resource for livelihoods, but their ability to influence government decisions on the use and management of resources is minimal. Resource user groups that are organised in the form of committees (e.g., Okavango Fishermen Committee) stand a better chance of acquiring space to negotiate for a better stake in the policy-making process, if given such a chance by policy planners (such as during the ODMP process). Discussions with local users at the resource level reveals that even after the conclusion of the ODMP planning process, local resource users see little or no change in the way things work, and in some cases, even experience increased marginalisation from accessing the resource, mainly as a result of policy decisions. As the outgoing chairperson of the Village Development Committee (VDC) of Seronga Village stated, "the ODMP was a short consultation and management process, there are many of these running parallel to each other and many of these end up introducing more restrictions on natural resource use".

The main issues that were raised during the ODMP stakeholder consultations and during the 2008–2009 fieldwork are discussed below. A number of potential pilot projects to test management options were identified during the ODMP. These would address some of the issues and concerns raised during the consultation process. However, the ODMP project could not fund any of these pilot projects, except for a clearing of small channels to improve access to the river in one part of the Ramsar site (NG 24). The main issues raised by communities are discussed below.

Water quality: (The threat of pollution from sewage and solid waste mishandled by tourismindextourism operations in the Delta)

Testing activities to monitor the quality of water from Mohembo to the Makgadikgadi at the distal end of the Delta continue through the BIOKAVANGO Project: "Through this we are able to monitor the health of the river and, if need arises, respond appropriately" (Deputy Director, DEA). The BIOKAVANGO project is also working on strengthening the capacity of Tawana Land Boardto lead inspections of tourism facilities.

River management issues: (channel blockages)

Department of Water Affairs (DWA) has been mandated to carry out small-scale clearing activities in some parts of the Delta as a pilot exercise. Noting the success of this activity, a member of the ODMP project staff observed that "livelihoods can be integrated into natural resource management resource access has been improved just by clearing a few channels" (Deputy Director, DEA). If successful, the implementation of the ODMP could upscale this to other parts of the Delta.

Land use conflicts (e.g., subsistence vs. commercial, livestock vs. wildlife)

Research conducted by Darkoh and Mbaiwa (2009) reveals that the existence of veterinary fences for controlling livestock disease transmissions are increasingly marginalising livestock farmers from access to certain grazing areas and water sources, resulting in conflicts among livestock-owning communities. These conflicts continue; the solutions proposed by the ODMP Action Plan are to: "improve maintenance of veterinary fences; mount public awareness on livestock diseases; and assess feasibility of providing livestock watering points in sand veld areas" (ODMP 2008: 152–161). These plans are to be implemented by the Department of Animal Health and Production (DAHP) using mostly in-house resources.

Wildlife, livestock and people movement (fences)

Veterinary fences also cause conflicts between livestock farmers and the Department of Wildlife and National Parks (DWNP), as livestock that roams into the cattle-free zone part of the Delta are often killed by the DWNP to control diseases, especially foot and mouth disease (FMD), while wildlife that roam into communal areas are merely driven back into the Delta. As Darkoh and Mbaiwa (2009: 163) note, "for farmers struggling to build their breeding stock after the CBPP (cattle-lung disease) outbreak which led to the culling of all cattle in the Delta, such killings of cattle were intolerably traumatic". But as a senior veterinary officer pointed out, to put this into perspective, "FMD is a disease of economic importance, we cannot afford to be seen to be failing to contain it, this would hurt our place in the European Union beef market, so we have to take stringent measures in controlling the movement of livestock, and policy is that if they [cattle] roam beyond the buffalo fence, they have to be killed" (Interview, February 2009).

Human-wildlife conflicts (elephants, predators)

Communities have raised concerns over the losses they incur due to predation of livestock by wildlife, both in the dry-land grazing areas (mostly lions, cheetah, and leopards) and at the riverfront (crocodiles and recently, elephants). Communities complain that compensation from the DWNP is below the value of the livestock lost; the DWNP maintains that the money paid for property lost is not full compensation but merely a token. Another major problem for communities is the destruction of their field crops by wildlife, especially elephants and buffalo. As one farmer noted, “Farming competes with elephants, every year around March the elephants go down towards the river in search of water and along the way they raid our fields; one can lose a whole crop in one night” (Interview, September 2008). The Problem Animal Unit of the DWNP in Seronga indicates that their office gets up to 25 reports a day (with an additional 25 at the *Kgotla*/traditional leadership authority and that interactions and field raids increase between the December and March cropping season. The Government of Botswana has little control over the elephant population as they are protected under the CITES convention as ‘endangered species’ and therefore cannot be culled. Their population has increased significantly over the last decade and, as a result, the human-elephant conflicts have also increased. The DWNP has introduced a ‘chilli pepper strategy’ along the ‘Human-Elephant Conflict Area’ between Mohembo and Gudigwa on the eastern side of the Okavango Panhandle. Through this strategy farmers are encouraged to grow chilli pepper plants around their fields to distract elephants from entering the crop fields. However, not many farmers have adopted this strategy and some have in fact exited agriculture.

Citizen participation and benefits from tourism

Ownership and management of the tourism industry is dominated by foreign businesses, especially from South Africa (Mbaiwa 2007). Jobs accessible to locals are therefore menial, unskilled, low-paying jobs in the support services (cooking, cleaning, and guiding). The ODMP (2008: 80) acknowledges the low benefits obtained by locals from the tourism industry: “It is evident that Batswana are not meaningfully participating. . . hence not appreciating the benefits thereof.” The new (December 2007) CBNRM Policy seeks to improve citizen participation in tourism development, but structural, capacity and political constraints continue to hamper meaningful participation and benefit sharing for locals. As a senior government official observed: “There are too many politics around the CBNRM, it has so far failed a lot of communities” (Interview, April 2009).

Conflicts in fishing

A senior fisheries officer in the DWNP District Office in Maun stated in an interview (April 2009) that “the shrinkage of fishing grounds due to droughts and poor floods in the Panhandle resulted in concentration of fishing efforts in one area and therefore conflicts arose. This called for regulation because these kinds of problems could have negative impact on the stock.” As shown in the stakeholder matrix provided earlier in this chapter, Fish Protection Regulations were introduced in May 2008. This was the main intervention introduced, partly as a result of the ODMP process and also because of the need for regulation. According to another senior fisheries officer, “For a long time the fishing sector remained unregulated, and therefore the time had come to do so” (Interview, April 2009). However, conflicts continue between the main fishing sectors (small-scale commercial and recreational). Small-scale fishermen blame recreational (angling and sport) fishing and the tourism industry in general for their decreased access to fishing grounds, especially those regarded as traditional fishing grounds by local fishermen. These fishermen view the regulations as unfair, discriminatory and punitive towards the small-scale commercial fishing sector. As one commercial fisherman in Seronga noted during a focus group discussion with commercial fishermen, "The main problem we have in the fishing sector is with big engine boats [used for recreation by the tourism industry]. These have not been prohibited as we requested and were promised. They continue to operate and we are being restricted. The Government favours them [the tourism sector] over us because they make the Government rich" (September 2008). In an interview in April 2009, the Head of the Fisheries Unit in Gaborone also noted that "commercial fishermen are mostly the ones affected by these [regulations] as they control the way they fish", but he stated that the new regulations actually favour locals as they only pay P200[1] for a three-year permit while lodges/ tour operators pay P3000[2] per annum in order to fish (catch and release method). However, anglers and other recreational fishers pay far more money to the tour operators for these activities, as part of the tour packages.

Land allocation issues and conflicts

The majority of land allocation conflicts arise from what the locals perceive as discrimination between locals and the white, foreign-owned tourism industry. Locals complain that they cannot access land for *molapo* (flood-recession agriculture) due to the ODMP 200m rule, which prohibits developments within 200 meters of the river, while this rule does not seem to be applied to the tourism industry. Interviews with the Land Board in Seronga reveal that it is increasingly difficult to apply this rule when allocating land for lodges and camps in the Delta due to political interference from the Department of Tourism.

Local tourism industry

The local tourism industry is the largest single (non-community) user of the resources of the Okavango Delta. As Mbaiwa and Darkoh (2006) state, "The rich wildlife diversity and scenic beauty of the Okavango Delta has therefore in the last decade made it the most important tourism industry destination in the country for both consumptive and non-consumptive wildlife utilisation" (2006: 02). The ODMP consultation process and subsequent research revealed evidence of the tourism industry receiving preferential treatment with regard to land use (see Bendsen 2005; and 2008/2009 fieldwork conducted by one of the authors). As noted by one local, "The Land Board is full of discrimination, they only allocate riverfront properties to white people" (Interview, Seronga, 2008). Racism

1 Equivalent to 23USD (September, 2013)

2 Equivalent to 354USD (September, 2013)

and elitism have also been cited as the main obstacles to local communities participating and benefiting from the tourism industry (see also Darkoh and Mbaiwa 2009 for a detailed discussion of racism in Botswana's tourism industry).

National government

As the custodians of the resource(s), the Botswana Government has, since independence in 1966, introduced and implemented a number of policies and regulatory instruments for use and management of the Okavango Delta. These include, for example, the Tourism Policy of 1990, the Wildlife Conservation and National Parks Act of 1986, the 1975 Grazing Land Policy, the 2007 CBRM Policy, and the 2008 Fish Protection Regulations. Several other strategies and programs are in place in the Okavango Delta, initiated and implemented by central Government institutions and local authorities.

Conservation of biodiversity (especially wildlife) is concentrated in the Okavango Delta region of the country, as is the lucrative wildlife-based tourism industry. About 8% of the Ngamiland District has been declared 'Protected Areas'; Moremi Game Reserve is the main one in the District, at 4,610 km^2 (Vanderpost 2007), having been expanded twice in 1978 and 1992. Wildlife Management Areas (WMAs) occupy another 40% of the Ngamiland region. In this region, other land use activities such as arable and livestock agriculture and communal settlements are competing for land, mainly with conservation and tourism, which are somewhat inseparable. What is clear is that the tourism industry – and therefore the tourism institutions of the government and private sector – has most of the control over the Okavango Delta due to the importance of tourism in this area for the national economy. Fenced, commercial ranching areas occupy about 10% of the Ngamiland District (Vanderpost 2007), which has exacerbated land use conflicts in the area, squeezing poorer communities who cannot afford to own cattle ranches into marginal areas. The Government's decision to declare the Delta a cattle-free zone and construct fences to keep cattle out resulted in the further exclusion of local communities from accessing the Delta resources (see Magole and Magole 2009; Mbaiwa and Mbaiwa 2003). As Magole and Magole (2009) note, in terms of management practices in the Okavango Delta, Government has taken all decisions after very little consultation with local communities dependent on the resource.

International community

International opposition to Botswana's early 1990s plans to develop the water resources of the Okavango Delta (mainly for water supply) marked the beginning of the loss of control of the Delta by the Government of Botswana, and subsequently residents of the Delta. International opposition to the Southern Okavango Integrated Water Development Project (SOIWDP) was driven by conservation organisations, specifically The World Conservation Union (IUCN) and Greenpeace.

Debates around the Delta developments were framed as having major 'environmental costs'; "without the project a lower cost water supply may be found for Maun" and "development prospects of the Okavango region through tourism may be better and certainly no worse..." (IUCN 1992: 133; see also Thomas and Selolwane 1998, for a detailed discussion of the SOIWDP opposition). While we are not suggesting that the SOIWDP would not have been potentially damaging to the ecological functioning of the Okavango Delta system, we argue that current use of the Okavango Delta, which has since shifted heavily towards tourism and conservation, does not benefit the local communities and, if allowed to continue unchanged, could even have damaging long-term social and environmental impacts. Over time, the tourism and conservation focus has transferred control of the Okavango Delta from the local communities, to the State (through the creation of the Moremi Game Reserve, and accompanying land grabs) and resulted in the subsequent loss of control to white, foreign-owned tourism and international organisation-dominated conservation programmes.

The philosophy, management approaches, projects and discourse on the use of the Okavango Delta increasingly reflect global environmental discourse and language – 'ecological threats', 'biodiversity of global significance', 'biodiversity conservation', 'international resource', to mention but a few – which frame the Delta as a resource under threat from human impact and needing to be 'protected'. This has gradually facilitated a process of internationalisation of a local resource, resulting in very little control by local communities, increasing marginalisation from the resource and no meaningful participation over decisions about its use. National Government policies and programmes use the same 'international' language. This thinking and framing of the resource has been further entrenched by the signing and ratification of international environmental conservation agreements and treaties, notably the Convention on International Trade in Endangered Species of Wild Fauna and Flora (CITES); the United Nations Convention on Biological Diversity; and the Ramsar Convention on Wetlands of International Importance, which culminated in the Okavango Delta Management Plan (ODMP). The Ramsar Planning Guidelines played a central role in framing and guiding the thinking during the ODMP process. The use of an ecosystem approach is central to the Ramsar thinking and management approach and is guided by such international scientific institutions as the Millennium Ecosystem Assessment which also advise institutions such as the Convention on Biological Diversity (CBD).

Discussion and conclusion

Despite the spatial coverage of the Delta communities in the ODMP consultation process, analysis of its products and local perceptions, coupled with current poor (if any) implementation of the plan, leads us to conclude that nothing has changed in the political ecology of the Okavango Delta, particularly for 'primary stakeholders'. ODMP Plans (i.e., Action Plans, Implementation Strategy and Monitoring and Evaluation Plans) have not changed the status quo, give no decision-making power or authority to primary stakeholders, and in fact, continue to disregard their traditional indigenous knowledge and Common Property Resource (CPR) management institutions. This was demonstrated in the new Fish Protection Regulations of May 2008, which present local fishing communities as a potential threat to the resource and institute measures to control how the resource is accessed and used, especially by local small-scale commercial fishermen. The new regulations assume that the CPR mechanisms of the local communities have failed or are themselves a threat to the resource. No civil society organisation exists in Botswana to represent the interests of the communities in decision-making about the use and management of the Okavango Delta.

Government (especially central Government) institutions involved in planning during the ODMP have proposed few or no new interventions as part of the new management plan. Equally, no new resources (technical or financial) have been made available for implementation of the plan. So in essence the different resource management sectors continue to function as they always did, before the inception of the ODMP, with little collaboration between sectors or appreciation of the integrated planning approach promoted by the ODMP.

As discussed in the previous section, the international community has a large stake in the use of the Okavango Delta through the tourism industry and conservation. This group of stakeholders has lost none of its power or control over the Delta, as they control the knowledge about how it 'should' be used.

In conclusion, therefore, the authors argue that, despite the extensive 'primary' stakeholder consultation process of the ODMP, ironically it was 'secondary' and 'tertiary' stakeholders of the Okavango Delta who were 'heard' in the stakeholder consultation and planning process. Perhaps accidentally, the ODMP in fact revealed the real political ecology of the Okavango Delta as 'part of larger gendered, classed, and raced struggles', where "increasing scarcities produced through resource enclosure or appropriation by state authorities, private firms, or social elites accelerate conflict between groups…" (Robbins 2004: 14).

Until the power dynamics of resource access, use and management at all levels (local, regional and international) can be adequately acknowledged and deconstructed, and environmental change and ecological conditions be seen for what they are as products of political processes, outcomes for the vulnerable and poor resource users will continue to deteriorate, posing a threat to the sustainability of their livelihoods and the resource base.

References

Bryant, R., and Bailey, S. (1997). *Third World Political Ecology*, Routledge, London and New York.

Darkoh, M.B.K. and Mbaiwa, J.E. (2009). "Land-use and resource conflicts in the Okavango Delta, Botswana" *African Journal of Ecology* 47 (Suppl. 1), 161–165.

Department of Environmental Affairs (2008). Okavango Delta Management Plan. Ministry of Environment, Wildlife and Tourism, Gaborone.

Forsyth, T. (2003). *Critical Political Ecology. The Politics of Environmental Science,* Routledge, Abingdon.

Government of Botswana. (2007). Community Based Natural Resources Policy Guidelines, Gaborone.

Hajer, M.A. (1995). *The Politics of Environmental Discourse: Ecological Modernization and the Policy Process*, Clarendon Press, Oxford.

Kgomotso, P.K., and Swatuk, L.A. (2006). "Access to water and related resources in Ngamiland, Botswana: Toward a more critical perspective and sustainable approach" *Physics and Chemistry of the Earth* 31, 659–668.

Magole, L. (2008). "The Feasibility of implementing an integrated management plan of the Okavango Delta, Botswana", *Physics and Chemistry of the Earth* 33, 906–912.

Magole, L and Magole, I.L. (2009). "The Okavango: Whose delta is it?" *Physics and Chemistry of the Earth* 34: 874–880.

Magole, L., Mzvimavi, D., and Mostumi, S. (2008). *Operationalising IWRMwithin the Okavango Delta, Botswana.* Proceedings of the Second IASTED Africa Conference, September 8–10, 2008, Gaborone, Botswana.

Mbaiwa, J.E., and Mbaiwa, O.I. (2003). "The Effects of Veterinary Fences on Wildlife Populations in Okavango Delta, Botswana," *International Journal of Wilderness,* Vol. 12, No. 3, 17–23.

Mbaiwa, J.E. (2009).*Tourism Development in the Okavango Delta,* Fact Sheet 8, Harry Oppenheimer Okavango Research Centre, Maun, Botswana.

ODMP Secretariat. (2005). ODMP Inception Report, Maun, Botswana.

Peet, R. and Watts, M. (2004). *Liberation Ecologies: Environment, Development, Social Movements,* 2nd Ed, Routledge, London.

Ramsar Convention Secretariat. (2007). *Managing wetlands: Frameworks for Managing Wetlands of International Importance and other wetlandsites.* Ramsar handbooks for the wise use of wetlands, 3rd edition, Vol. 16. Ramsar Convention Secretariat, Gland, Switzerland.

Ribot, J.C, and Peluso, N.L. (2003). A Theory of Access, *Rural Sociology* 68 (2), 153–181, Rural Sociology Society.

Robbins, P. (2004). *Political Ecology: A Critical Introduction,* Malden, Oxford and Carlton: Blackwell.

Scudder, T., Manley, R.E., Coley, R.W., Davis, R.K., Green, J., Howard, G.W., Lawry, S.W., Martz, D., Rogers, P.P., Taylor, A.R.D., Turner, S.D., White, G.F. and E.P. Wright. (1993). The IUCN *Review of the Southern Okavango Integrated Water Development Project*, IUCN, Gland, Switzerland.

Thomas, A. and Selolwane, O. (1998). "A Forced Policy Change? The Campaign Against the Southern Okavango Integrated Water Development Project (SOIWDP)." Development Policy and Practice Research Group Working Paper 42, GECOU Working Paper 11, The Open University.

Vanderpost, C. (2007). "Protected areas in Ngamiland, Botswana: investigating options for conservation-development through human footprint mapping", *International Journal of Environmental Studies* Vol. 64, no. 5, 555–570.

Watts, M. (2000). "Political Ecology", in E. Sheppard and T.J. Barnes (eds) *A Companion to Economic Geography*, Blackwell Publishing, Malden.

Wester, P., Shah, T. and Merrey, D.J. (2005). "Providing Irrigation Services in Water-scarce Basins: Representation and Support", in *Irrigation and River Basin Management: Options for Governanceand Institutions*, Cambridge, Mass: CABI.

3 The Role of Universities in Community Development and Empowerment: The Case of the Okavango Delta Management Plan

Lapologang MAGOLE and Rachel DEMOTTS

"We are happy to see the representatives of the University of Botswana here. They do not visit us often." *Resident of the Okavango Delta*

Abstract

As natural resource management (NRM) becomes more and more participatory, it can also become more contentious. Rather than remaining outside of such processes as observers, universities have a clear responsibility to assist local residents and support the development of effective tools for local participation in projects that will have profound effects on local communities and livelihoods. To illustrate the importance of this role, we discuss the role of the then Harry Oppenheimer Okavango Research Centre (HOORC) in the development of the Okavango Delta Management Plan (ODMP). In its role as the primary facilitator of community participation, HOORC worked to ensure that throughout the planning process, the voice of local residents was heard. HOORC also prioritised providing information about the planning process so that local residents could participate in the creation and implementation of the plan and so that the plan would address both their interests and concerns. We also address the difficulties experienced in this process and draw lessons for university involvement in environmental consultation in rural areas that are increasingly affected by tourism. While not without challenges, HOORC was able to facilitate community empowerment and effective participation in the ODMP planning process.

Introduction

Until perhaps two decades ago, Maun was a dusty, one-road-in-and-out kind of wayside with a small airstrip and a hotel on the edge of a vast wetland teeming with wildlife. But the explosion of interest in the Okavango Delta as one of the premiere safari destinations in Africa has led to the almost haphazard expansion of the town. Air-conditioned hotels with turquoise swimming pools sit just behind concrete and razor wire fences, insulated from neighbouring mud huts and women washing clothing in the river while Land Rovers cruise around donkey carts and trotting goatherds on the main road. Maun now stretches out along the main road south of the Delta, built on Kgalagdi sand and western dreams of the wild and wildlife in rural Africa. This growth has created not only hotels and tourist facilities, but walls between expatriates and tourists on the one side, and local residents on the other. The remaining commonality, however, is mutual dependence on natural resources – whether for subsistence, water, and fishing, or for stunning views of elephants from rustic lodges overlooking scenic water holes. Consequently, the equitable and effective management of these resources is perhaps more contentious and crucial than ever before.

Participatory management of NR is now an expectation in most of Africa. Fortress-style protected areas and fenced national parks, for example, have been heavily criticized for their exclusion of local residents – either through forced displacement or marginalization outside the fence (Brockington 2002; Chatty and Colchester 2002; Fabricius and de Wet 2002; MacKenzie 1988; Neumann 1998, 2001). But while community participation is the language of the day in conservation, there are growing questions about how well it actually works (Campbell and Vainio-Mattila 2003; Coffman 2003; Songorwa 1999; Swatuk 2005). Communities are clearly not on equal footing with Government agencies, NGOs, scientists, researchers, and others who advocate and implement conservation, and they often lack the power and capacity to be heard or make demands. But as the primary users of natural resources, they often have the most to gain – or lose – from environmental planning processes and natural resource management.

In the Delta, local residents retain rights to use land and water in some areas, while in others these rights are increasingly curtailed (Swatuk, 2005). Community trusts establish local rights to create community-based tourism enterprises which often rely on safari hunting for income, but a new national policy may soon require 65% of all project income to be given to the national Government[1] – creating fears that community-based conservation is dead, and reflecting additional pressures on community-based conservation projects to find ways to make up for this lost income. In the Sankuyo Tshwaragano Management Trust on the edge of the Delta, for example, BWP 1.8million (approximately USD$285,000) is generated annually from safari hunting alone, which excludes income from the Santawani Lodge and the Kazikini Safari Camp.[2] This possible change reflects the extent to which community participation and input is essential for the creation of inclusive and fair conservation and environmental management policies.

In an area such as the Delta with a growing tourism industry that also faces mounting pressure from Government, how can the voice of local residents be heard and respected in planning processes? We argue in this chapter that in some cases, universities can address this gap. University relationships with local communities can engender opportunities to assist residents in making their concerns

1 N. Ndlovu, "New policy may lead to job losses," *Mmegi,* 14 February 2007. Available at http://www.mmegi.bw/2007/February/Wednesday14/59.php

2 Ibid.

heard and having them addressed in a substantive way. This kind of process, however, requires a willingness on the part of universities to make participation and consultation more important than data-gathering for pure research purposes – something which is not only difficult to do, but that is also often without incentive for faculty.

In this chapter, we will discuss a unique process in which the University of Botswana (UB) was called upon to consult with local communities for the purpose of designing an environmental management plan for the Okavango Delta. The University's research centre in the Delta, the then HOORC and now Okavango Research Institute (ORI), allowed locally-based researchers and faculty to conduct consultations and solicit local views while using their own experiences and expertise in the immediate area – contrasting with the usual approach of relying exclusively on consultants to write reports on local areas with which they may or may not be very familiar.[3] To illustrate this process, we will first discuss briefly some conceptual issues in community-university relationships with an eye on particular challenges for African universities. Second, we describe the creation of HOORC and how it was linked to the Okavango Delta Management Plan (ODMP) planning process. Third, we highlight community perspectives on the ODMP itself and the consultation process, and finally we offer some thoughts on learning from this process of interaction.

On equal footing?

Universities play many roles in relation to the communities in which they reside. With the recognition that conducting research without making a contribution in return can easily become extractive, universities are increasingly trying to form partnerships and mutually beneficial relationships with local residents, realising that communities do not always gain from university research programmes and that key stakeholders in public planning are often excluded (Lederer and Seasons 2005). This ranges from trying to bring community members to the university for events and to take advantage of university resources (Bruning *et al.* 2006) to understanding better the importance of interdisciplinary approaches among faculty members involved in community projects (Amey and Brown 2005). Research approaches such as participatory rural appraisal (PRA) are particularly associated with Freire (1970) and turn on the notion that people are not subjects, but rather need to be full participants empowered to act of their own volition. In this vein,

[3] Consultants were also employed in this process, but they were based in Gaborone (the capital of Botswana) and/or as far afield as Cape Town, South Africa – more than 2,000 kilometers away. While not uncommon, this difference can have a significant impact on the quality and insight of local consultation. Moreover, consultants hired by Government agencies often struggle to emerge as community advocates because of their own financial interests and the desire not to 'rock the boat' by recommending approaches that may not be popular with Government officials or critiquing existing policies.

more broadly, conducting research in a participatory fashion and sharing results with local participants – particularly in developing countries – has also become a priority (Cernea 1991; Chambers 1983, 1997).

In some cases, researchers have turned to action research strategies which facilitate not only collecting information but also the actual design of research in concert with local stakeholders so that it specifically addresses local planning or development needs and is directed at social change and empowerment (Small and Uttal 2005; Williams *et al.* 2005). This focus on increased participation and the importance of local input at every step – not just during data gathering – reflects a growing realisation that the assumption that effective labour alone could spark community development misunderstood or underestimated the dynamics of community change processes (Williams *et al.* 2005). Communities need not only knowledge, but capacity and empowerment to effect change. For universities to contribute to change, however, requires a long-term commitment to forming lasting local partnerships; without it academics may easily collect data but remain *irrelevant* (Stoecker 1999) to larger social and political processes.

Many community-university partnerships seem to revolve around issues such as urban planning, public health, and community development. Involving local residents in environmental management, however, seems to be less common, perhaps due to the perceived need of technical expertise as the primary tool for effective projects. In Africa in particular, for example, communities living near national parks are often marginalised from participating in management because they are said to lack the necessary knowledge. This argument has two fundamental flaws. First, it ignores the need to assess and incorporate local knowledge of the natural environment in any planning; and second, it discounts the important role that local residents already play – and must play – on a daily basis in managing NR if any future efforts are to be successful. Local residents are, after all, just that – *those people who live with and utilise natural resources in the immediate area on a daily basis.*

But even as local participation is crucial in environmental management and conservation, it is contentious and difficult to facilitate. Again, in the case of Botswana (and rural Africa more generally), "Different stakeholders have very different views on the levels of participation taking place in particular projects. Furthermore local people find it difficult to voice their concerns about the environment and issues of sustainability given the power relations involved in this 'participatory' process" (Twyman 2000: 323). This suggests that it is not only the substantive issue at stake that matters, but also who is talking and who is listening in the dialogue. The highly politicised nature of conservation in rural Africa makes it that much more crucial that participation and consultation are conducted inclusively and openly. Given local experiences with Government environmental agencies as primarily interested in disciplining local behaviour, universities may be better positioned to talk with local residents – especially when they are also members of the local community who are invested in equitable conservation measures.

The Harry Oppenheimer Okavango ResearchCentre (HOORC)

The establishment of the University of Botswana's Harry Oppenheimer Okavango Research Centre (HOORC)[4] in Maun, Ngamiland District, Botswana – a thousand kilometres away from the main campus in the capital city of Gaborone – represents not only a commitment by the university to advance research on natural resources, but an opportunity to work more closely with local communities in a rural setting that is increasingly a locus of tourism and international conservation and development initiatives. In 1991, the Department of Water Affairs (DWA) attempted to initiate a controversial dredging program in the Delta that had the potential to substantially alter its waterways and access points. Local and international objections were raised, from residents of the area as well as international environmental organisations such as Greenpeace. Ultimately, the Government of Botswana dropped the project, and one of the important outcomes of the opposition was the "Government's realisation that it lacked sufficient background knowledge and an adequate scientific base to formulate policy and management decisions on the Delta. Government also realised that an impartial body was needed to monitor developments in the Delta."[5] In this way, ORI's beginnings are intimately linked to the need for better local participation and meaningful consultation in the management of the Delta.

Currently, ORI consists of approximately 20 faculty members from several disciplines who work on the defined research niche area of wetlands management through studies on Ecosystems, Governance, Land and Water, Livelihoods, and Tourism. Researchers include both natural and social scientists. The researchers work collaboratively on many projects, ranging from biodiversity conservation and wetlands ecology to trans-boundary environmental governance and the local impacts of tourism on livelihoods. ORI is also involved in regional and international partnerships with universities in Southern Africa as well as Europe and the United States (and is exploring links to China and Japan). So while the Institute is clearly grounded in work on issues of importance in and around the Delta, it is also committed to the creation of partnerships locally and globally. ORI runs its own MPhil/PhD programme in Natural Resources Management, in part to increase opportunities for students from Botswana to study and conduct important research on environmental issues.

---

4 Since a strategic planning exercise in 2010, the Centre has been transformed into a wetlands research institute, renamed the Okavango Research Institute (ORI). HOORC will continue to be used in this paper largely to refer to historical events.

5 Origins of HOORC/ORI, described at http://www.ori.ub.bw/.

Reflecting its original purpose, ORI continues to make outreach a priority for more than just research purposes. Outreach focuses on participation, education and capacity building, underpinned with research on issues of practical importance to the conservation of the Delta. Communities are a key part of this because "ultimately the conservation of the Okavango rests with them".[6] The commitment to working with communities extends to making research useful and more accessible as, "traditionally, scientists have been blamed for failure to generate meaningful dialogue with different lay-audience groupings such as policy makers and local communities. This has partly been blamed on the complex nature in which ecological issues are presented making it very difficult to be understood by non-scientists or non-technical experts."[7] The importance of communication is clear – not only in sharing traditional research projects, but also through engendering interdisciplinary research that takes into account local knowledge and scientific approaches. This priority made HOORC/ORI an obvious, important partner in the ODMP process.

Linking people and policies: The Okavango Delta Management Plan

There is a general consensus among scientists and practitioners that the Okavango Delta is in nearly pristine condition (Turton *et al.* 2003). The potential for increased use of the Delta, however, calls for systematic management of the entire basin and the Delta in particular. Pressures are caused by tourism (lodges, airplanes, boats, supplies moving in and out) and, at times, the poverty of the communities residing in the area that rely on fishing and agriculture for subsistence. Developing the ODMP represents Botswana's effort to contribute to what, it is hoped, will eventually become a basin-wide management strategy.

The ODMP, initiated by the Government of Botswana, seeks to integrate water and land management focused on the Delta and the communities residing within it. The ODMP focuses on the 6,684,000 hectares of the Delta that were declared a Ramsar site in 1996. To fulfil the commitment made in signing the Ramsar Convention, Botswana must put a management plan in place; Article 3.1 of the Convention states that, "The Contracting Parties shall formulate and implement their planning so as to promote the conservation of the wetlands included in the List, and as far as possible the wise use of wetlands in their territory."[8] This plan-

6 ORI webpage, http://www.ori.ub.bw/, Outreach page.

7 ORI Outreach Project, "Effectiveness of Science/Environmental Communication Strategies Among Delta Communities," O.T. Thakadu, ORI, and Dr. O. S. Tau, Centre for Continuing Education, University of Botswana. Full project abstract available at http://www.ori.ub.bw/, Ongoing Research Projects page.

8 Convention on Wetlands of International Importance especially as Waterfowl Habitat. Ramsar (Iran), 2 February 1971. UN Treaty Series No. 14583.As amended by the Paris Protocol, 3 December 1982, and Regina Amendments, 28 May 1987.Available online at http://www.ramsar.org/index_very_key_docs.htm.

ning initiative also falls within national priorities and the ideal of sustainable development set out in National Development Plan 9 and captured in the country's Vision 2016, which singled out the Okavango Delta area as a candidate for such a development strategy and thus, required planning to promote prudent use of its resources to benefit the local communities, the nation and the international community at large. As a major tourist destination, the Delta is also important to the national objective of economic diversification.

Within the Delta, conflicts have emerged between and among the layers of stakeholders with vested interests in how the Delta is managed and who is able to use resources. There is, for example, competition between recreational, commercial and subsistence fishing; between local residents and tour operators; within and between Government authorities such as the Department of National Wildlife and Parks and the Department of Animal Health and Production – and of course, among these various groups – on issues of subsistence, use, and economic benefit. The ODMP process sought consensus among all user groups and responsible institutions on resource allocation, use, and management to help resolve existing conflicts and pre-empt potential conflicts. This situation, combined with a more general need to conserve natural resources in the Delta, called for a much more creative planning strategy than adopted in conventional approaches.

In a move that was somewhat non-traditional for Botswana's planning system – which tends to be centralised, hierarchical, and Government-controlled – UB through HOORC was involved from the outset in the ODMP, providing initial expertise for development of the plan proposal. But most important for this chapter, HOORC was tasked with mobilising community participation in the process and advising the project Secretariat on stakeholder involvement in general. To facilitate stakeholder participation, the ODMP team (including participatory planning specialists from HOORC) performed a stakeholder analysis premised on the broad definition of a stakeholder as "any interested or affected individual or group" (ODMP 2005b: 25). The analysis identified local communities as the only group of stakeholders who had a 'very high' need for capacity building to be able to participate effectively in the process.

This analysis also placed the communities resident inside the Delta at the centre of the stakeholder map, making them the only primary stakeholders in the process. Secondary stakeholders reached into communities residing in the rest of the basin, and only at the level of tertiary stakeholders did government and other outside organisations take their place. Outside of these three groups, there were also key stakeholders, a group which although not dependent on the Delta resources possesses the potential to influence its day-to-day management – for example, international actors such as the Ramsar Bureau and the Permanent Okavango River Basin Water Commission (OKACOM). This mapping of interests that puts communities at the heart was a crucial early step in the process, because a community focus then became the thrust of the ODMP's participatory planning process –which was assigned to HOORC.

The approach adopted was guided by a threefold set of principles: *Communicate – Integrate – Manage*. These principles are further elaborated as follows:

- Planning with all levels of society, by all with a responsibility to manage and for all interest groups;
- Confronting the challenge of integrated planning by embracing ecosystem management approach to address the socio-economic challenges with ecological and bio-physical aspects; and
- Elevating the livelihoods elements in natural resource management: Poverty reduction, HIV/AIDS prevention and gender equality (ODMP 2005a).

At the same time, the ODMP Secretariat – housed by the Department of Environmental Affairs – had a great deal of latitude over project resources and procedures. A range of planning topics from veterinary and disease issues to fisheries and tourism were divided into components and assigned to appropriate agencies and organisations for preparation.

Component 3 was HOORC's primary task: UB-HOORC – Research, Data Management, and Participatory Planning

- Set up a comprehensive database, Okavango Delta Information System (ODIS) and facilitate its access and accessibility by all stakeholders including individuals and community groups
- Build up a library with material relevant to the ODMP planning process
- Make results of ongoing research activities available to the ODMP planning process
- Develop an ODMP-specific research strategy to fill data and information gaps
- Provide participatory planning expertise, thus to monitor and advise on the participation of stakeholders in the ODMP planning process, and in particular to mobilise and support the primary stakeholders to participate (ODMP 2005b).

The goal of the participatory planning part of the component was to drive the project towards achieving, "the creation and strengthening of ownership" of the plan by all stakeholders, especially communities living with and using the resources within the Ramsar site, in order to facilitate acceptance and hence implementation (Government of Botswana 2002).

Community voices, university advocates

Together with the ODMP Secretariat, HOORC planners drew up a consultation and participation strategy based on the theme of "Embracing community power, community knowledge and community action" (Magole 2005). Consultations began in 2004 and continued until late 2006, often in the form of a *kgotla*, or traditional village meeting and public forum for discussing issues. Communities were clearly struggling with the effects of poverty, a lack of education, and political marginalisation, all of which had a disempowering effect on community members. Moreover, as the consultation process began, communities were often hostile and did not trust the planning 'machinery,' which they initially associated with a Government they considered to have neglected them for years. There was, however, both openness to – and disappointment in – the HOORC representatives in the initial discussions, as indicated by the quote at the beginning of this chapter: "We are happy to see representatives of the University of Botswana here. They do not visit us often" (Bendsen 2005b: 1).

Early in the process, community members articulated many concerns, which coalesced around three central issues. First, they were frustrated by the knowledge that the ODMP process was already underway and that they had not been involved from the very beginning, which influenced the value they perceived would be attached to their input. As one resident commented, "You have already progressed so far with the ODMP project without consulting the communities. We doubt that our comments will be incorporated now" (Bendsen 2005b: 1). This identifies a crucial issue – that while participation was always seen as part of the ODMP process, it was not embedded in the beginning stages – for example, by incorporating local representation into government, consultant, and NGO discussions of implementation.

Second, communities talked about 'participation fatigue,' as they are often asked for ideas and input but rarely see tangible outcomes from these efforts: "We are tired of all the meetings, as solutions are not implemented" (Bendsen 2005b: 5). Often, consultation is just that – asking for input, but without any guarantee that such input will be considered or will affect the project outcome. Third, some community members expressed the desire to be treated as equals in the ODMP and to work cooperatively with those who had come to consult them:

> Consultation has taken place in other instances but what is lacking is that we work together and co-operate. We will only believe that we are given a true chance to participate when our concerns, aspirations and needs are taken into consideration and are addressed (Bendsen 2005b: 3).

HOORC Faculty members compiled and analysed comments after the first round of *kgotla* meetings, and then took them back to community and resource user group leaders in a workshop where the issues were confirmed and ranked. Another round of *kgotla* meetings followed, and this time, community members began to make requests, substantive suggestions, and even demand certain actions. One community member told planners that they should go away and leave time for the community to discuss the ODMP on its own before expecting them to share ideas and thoughts in the formal planning process. Participants also recognised their need for capacity building:

> We need educational workshops to improve our participation and understanding and benefit from the project. We need education so that after three years of planning we become part of the management process (Bendsen 2005b: 7).

Others suggested that there was a need for a local structure to communicate with Government and others on important issues such as the ODMP, citing the fact that they often get conflicting information about the same issues from different Government departments. However, some began to view the ODMP process more optimistically:

> This is an opportunity to have our concerns and suggestions included in the plan. We are thankful that Government has changed its approach and is acknowledging our expertise and gives us a chance to have an input before decisions are made. (Bendsen 2005b: 8)

But perhaps most importantly, community members began to assert their pride of place:

> We know better than all the educated people about what is happening in the Delta as we have lived here all our lives. We have a lot of traditional knowledge of resource use and conservation that we can share with you. We can advise you on the plan (Bendsen 2005b: 9).

A second workshop to consolidate comments proposed solutions, policies, resource use regulations, and institutional arrangements to local leaders for their comments, additions and advice. Throughout the process HOORC professionals advocated for and encouraged members of the community, and perhaps most importantly, changed the view of many community members that meetings were pointless and that their comments would not be heard and incorporated. This shift of ground away from the confrontational to cooperative opened up space for participation and discussion while building relationships between HOORC and the communities in and around the Delta. With increased knowledge and discussion, according to one local resident, "it is becoming clearer to us what the ODMP means" (Bendsen 2005b: 23).

During the last planning phase meetings, community members expressed appreciation for their involvement and even the planning approach itself. Many stated that it was the first time they had ever been consulted. But more than this, what emerged clearly was the need for quality, intensive consultation and ongoing efforts to build capacity. Reflecting both the desire to participate and the kind of local environmental knowledge that is crucial for the success of the ODMP, one local resident commented:

> We appreciate the idea of education. Some issues should be addressed through conservation awareness. We have to look at conservation of plants. Baobab trees have been depleted in the vicinity of our village (Bendsen 2005b: 26).

Communities are often consulted but never get feedback on their concerns, requests and suggestions. The platform for dialogue with local residents is thus often uneven, as they are usually consulted too late in a process or are expected to talk about foreign concepts which are not explained adequately. The engagement of the HOORC Faculty that had the time, interest and freedom to work with communities boosted the community participation profile of the ODMP. Faculty members took the time to explain proposals and concepts to the communities, providing information about the process and substantive issues. Discussions ranged widely – from procedural, technical, and substantive issues, to positive and negative feedback on all of these issues.

Furthermore, because of their independence, HOORC Faculty could be critical of Government agencies and other stakeholders for their failure to respond to community issues or queries. For example, at a workshop with community leaders and representatives held in Shakawe in 2006 (on the northern edge of the Delta), Government planning departments were asked to present their resource management proposals. In so doing, they were asked to address not only conservation, but also how these proposals would address community issues and how local residents could expect to benefit. During discussions, HOORC's representative at the meeting became increasingly aware that community members had concerns but were not stating them publicly. HOORC's planning specialist suggested to the community that they, in fact, had the ability to reject the proposals that were being made to them – a step which either had not occurred to local residents, or a step which they felt could not be taken. At this suggestion, the workshop participants rejected the DWA report, which had concentrated on hydrological modelling, as it did not adequately address addressing their issues with water resources management (sanitation) and access (channel blockage) and directed DWA to present them with another report. This advocacy pleased the communities and was an important step for their participation in the process.

Lessons and concluding thoughts

Communities such as those in the Okavango Delta have to deal with government regulations about their use of the natural environment every day. Livestock are monitored for disease, veterinary fences are built to keep livestock out of parks and wildlife areas, fishing practices are curtailed by tourism policies, subsistence hunting is turned into poaching, and the kinds of crops that can be grown depend on access to water and fertile soil. These regulations can create resentment, and affect peoples' ability to survive and prosper. So in this case, the opportunity to talk with outsiders who were *not* government on these issues was an important contributing factor in community willingness to participate in the process. Moreover, HOORC Faculty members with experience working in the area were able to enter communities knowing the lay of the land and with the ability to talk about environmental issues from an educated, teaching-centric perspective.

What was also important about HOORC's approach to the ODMP process was that it did not regard consultation as an isolated, one-off undertaking, but rather moved it through multiple iterations. Faculty members collected the views and thoughts of community members, elaborated on issues to help raise awareness, went away to discuss and synthesise inputs, returned to meet with local leaders, and so on. This demonstrated the commitment of HOORC and the Government (especially the DEA) to 'get it right', and to offer the maximum opportunity possible for local residents to share their thoughts, concerns, and criticisms. It also created a space in which to build trust – a crucial issue in an area that faces many kinds of conflicts emanating from competition for natural resource access and use.

One of the most important lessons, even in a research Institute founded in large part as a means for better communication and participation in the Delta, is reflected by the comment at the top of this paper which was made during the ODMP process: "We are happy to see the representatives of the University of Botswana here. They do not visit us often."[9] Even with ORI's clear commitment to the local area, it is a constant challenge to be fully inclusive and participatory all the time. Practicalities do get in the way; the Delta is vast and much of it is difficult to reach. Moreover, even with ORI's somewhat non-traditional purpose, its Faculty members are still required to publish in scientific, peer-reviewed journals whose review boards may not be terribly concerned with local participation.

It proved to be a challenge to convince Government officials who hold positions of power over communities and their resources to appreciate the need for genuine community involvement and partnership in resource management – to embrace community, knowledge, power and action. The entrenched culture of Government departments as the 'police' of resources will take time to uproot. It also proved difficult to encourage disempowered communities to assume their rightful place in resource management. After many years of being marginalised and ignored in natural resources governance, communities sometimes suffer from a culture of complacency and dependence on Government to solve problems. Communities have been taught – and learned well – that they are not valued in the process: as one participant stated, "In the past some policies have been imposed on us and I therefore appreciate the idea of consultation in the ODMP" (Bendsen 2005b: 24). But HOORC's combination of education and advocacy efforts began to lessen these effects.

The ODMP process clearly raised the standing of the University of Botswana in the community and made many more local residents aware of its presence as a growing institution – leading one local resident to comment that, "We would like

9 Resident of the Okavango Delta, ODMP Consultation Process, 2006. Interviews and data held at ORI, Maun, Botswana.

to see HOORC extended to a full-fledged university" (Bendsen 2005b: 25). ORI continues to try and find ways to incorporate participatory approaches in its own research, and to conduct outreach that is useful in its own right. The ODMP process lends important insights into the best ways to achieve this. In addition, even though the ODMP has now entered the implementation phase, ORI has pledged to remain involved for as long as it is needed. This is an important opportunity for ORI to build on its knowledge of the community and to nurture relationships with local residents, while becoming an important resource for those who live in and around the Okavango Delta.

References

Bendsen, H. (2005a). "Documentation of Second Round of Kgotla Meetings". Maun: ODMP/University of Botswana Harry Oppenheimer Okavango Research Centre.

__________ (2005b). "Observations, Suggestions and Concerns from Communities Regarding ODMP Activities Grouped According to Components." Maun: ODMP/University of Botswana Harry Oppenheimer Okavango ResearchCentre.

Brockington, D. (2002). *Fortress Conservation: The Preservation of the Mkomazi Game Reserve, Tanzania,* Oxford: James Currey.

Bruning, S. S. McGrew, and Cooper, M. (2006). "Town-gown relationships: Exploring university-community engagement from the perspective of community members," *Public Relations Review* 32: 125–130.

Campbell, L.M. and Vainio-Mattila, A. (2003). 'Participatory Development and Community-Based Conservation: Opportunities Missed for Lessons Learned?' *Human Ecology* 31(3): 417–437.

Cernea, M. (ed.). (1991). *Putting People First: Sociological Variables in Rural Development*, Oxford: Oxford University Press for the World Bank.

Chambers, R. (1983). *Rural Development: Putting the Last First,* Essex: Longman Scientific and Technical.

___________(1997). *Whose Reality Counts? Putting the First Last,* London: ITDG Publishing.

Chatty, D. and Colchester M. (eds.). (2002). *Conservation and Mobile Indigenous Peoples: Displacement, Forced Settlement, and Sustainable Development,* New York and Oxford: Berghan Books.

Coffman, J. (2003) 'Buying (Into) and Selling Conservation Among Maasai in Southern Kenya,' pp161–188, In B. I. Logan and W.G. Moseley (eds), *African Environment and Development: Rhetoric, Programs, Realities.* (2003) Aldershot: Ashgate Press.

Fabricius, C. and de Wet C. (2002). 'The Influence of Forced Removals and Land Restitution on Conservation in South Africa,' pp142–157, In D. Chatty and M. Colchester (eds) *Conservation and Mobile Indigenous Peoples: Displacement, Forced Settlement, and Sustainable Development,* New York and Oxford: Berghan Books.

Freire, P. (1970). *Pedagogy of the Oppressed,* London: Penguin Books.

Government of Botswana, National Conservation Strategy (Coordinating) Agency (2002) "ODMP project proposal."

Lederer, J. and Seasons, M. (2005). "Understanding the University-Community Alliance: The Waterloo Experience," *Canadian Journal of Urban Research* 14(2): 237–260.

MacKenzie, J.M. (1988).*The Empire of Nature: Hunting, Conservation, and British Imperialism*, Manchester, NY: Manchester University Press.

Magole, L. (2005). Report of the ODMP consultative workshop for community leaders and resource user group representatives, held at Shakawe Sub-LandBoard Chamber, 24th to 28th January 2005, Shakawe village, Botswana.

Neumann, R. (1998). *Imposing Wilderness: Struggles over Livelihood and Nature Preservation in Africa,* Berkeley: University of California Press.

______________ (2001). "Disciplining Peasants in Tanzania: From State Violence to Self- Surveillance in Wildlife Conservation", pp 305–327, In N. Peluso and M. Watts, (eds) *Violent Environments,* Ithaca: Cornell University Press.

Okavango Delta Management Plan (ODMP). (2005a). Publicity material articulating the project principles and rational.

Okavango Delta Management Plan (ODMP). (2005b). "Inception report, Volume One." (Main report).

Ramsar. (1971). The Ramsar Convention Manual. "A Guide to the convention on Wetlands." Gland, Switzerland.

Small, S. and Uttal, L. (2005). Action-Oriented Research: Strategies for Engaged Scholarship," *Journal of Marriage and Family* 67 (November): 936–948.

Songorwa, A.N. (1999). 'Community-based wildlife management in Tanzania: Are the Communities Interested?' *World Development* 27(12): 2061–2079.

Stoecker, R. (1999). "Are academics irrelevant? Roles for scholars in participatory research", *American Behavioural Scientist* 42: 834–848.

Swatuk, L. (2005). "From 'Project' to 'Context': Community Based Natural Resource Management in Botswana", *Global Environmental Politics* 5(3): 95–124.

Twyman, C. (2000). "Participatory conservation? Community-based Natural Resource Management in Botswana", *Geographical Journal* 166(4): 323–335.

Williams, A., Labonte. R., Randall, J. and Muhajarine, N. (2005). "Establishing and sustaining community-university partnerships: A case study of quality of life research", *Critical Public Health* 15(3): 291–302.

4 Participatory Democracy in a Fledgling Community Organisation – The Case of Tubu Joint Management Committee, Okavango Delta, Botswana

Lin CASSIDY, Lapologang MAGOLE and Bontekanye BOTUMILE

Abstract

In recent years, there is increasing appreciation of the importance of control of access to resources under common property regimes (CPRs) for the sake of sustainability of natural resources. The role of good governance in access to, and use of, resources is now understood as being critical to both social and ecological sustainability. In particular, governance that is based on even representation and open to community participation is seen as necessary for successful resources management. In many countries in southern Africa, the devolution of decision-making to include the local level has been problematic. While outside support is needed to help local leaders maintain good governance systems, this must be balanced against the institutional weaknesses resulting from the loss of local-level control due to the centralisation of national governments at independence. Recent research in the region supports the notion that good governance should be based on transparency, accountability and democratic participation in decision-making. In this chapter, we document the process of building sound governance practices in a fledgling community organisation tasked with managing the natural resources in its area. Household surveys were administered in the community asking about their views of this fledgling joint management committee (JMC). Research results show that while lessons were learned of the importance of transparency in the democratic process, further outreach is needed to explain the existence and role of the committee to most community members.

Introduction

In recent years, concern over the sustainability of natural resources use has seen a shift away from blaming communal tenure, or common property, systems (cf. Hardin 1968) as a major cause for misuse of natural resources. Increasingly, there is a more complex appreciation of the actual management or control of access to resources under such property regimes (Ostrom *et al.*1999; Dietz *et al.* 2003). The role of good governance in access to and use of resources is now understood as being critical to both social and ecological sustainability. In particular, governance that is based on even representation – and open to community participation – is seen as necessary for successful resources management (Anderies *et al.* 2006). In many countries in southern Africa, the devolution of decision-making to include the local level has been problematic. While outside support is needed to help local leaders maintain good governance systems, this must be balanced against the institutional weaknesses resulting from the loss of local-level control due to the centralisation of national governments at independence (Balint and Mashinya 2006; Chabwela and Haller 2010; Collomb *et al.* 2010).

In southern Africa, governance during the colonial era was dominated by 'command and control' approaches .(Holling and Meffe 1996). Over the past few decades, however, there have been attempts in Southern Africa to re-introduce local level management of access to natural resources, starting with the CAMPFIRE programme in Zimbabwe (Child and Peterson 1991). Yet most of these so-called communal approaches to natural resources management have been inherently flawed, in that they have been imposed from the top down (Murphree 2004), and have not necessarily resulted from a re-galvanisation of collective action driven by local interests (Adhikari 2005; Mutandwa and Gadzirayi 2007).

The Botswana attempt to avert the problems associated with open access to natural resources, particularly of economically important wildlife, was based on this kind of top-down approach. In 1995, the Botswana government introduced a community-based natural resource management (CBNRM) programme in key villages situated inside the wildlife management areas in northern Botswana, with limited success (Blaikie 2006). In Botswana CBNRM is based on the principle of creating incentives for rural communities to support the presence of wildlife in their areas .(Government of Botswana 2007). In an area where wildlife-based tourism is a major revenue earner, the potential for local people to benefit is high. However, one of the biggest challenges to the programme's success is that the benefits tend to accrue at the community level, while the costs (e.g., livestock losses, crop raiding) are still borne by individual households – an issue addressed in CAMPFIRE in Zimbabwe (Child 1996), but not in Botswana. The focus on wildlife as the main, high-earning resource has tended to detract from proper management of other key resources that people rely on, such as water, thatching grass, grazing, and other food, building and medicinal plants (Shackleton, Shackleton *et al.* 2001). Further, in traditional societies, wildlife is seen as the men's resource, and women often become marginalised in decision-making, even where they may have an important role in hunting small game or birds (Hunter *et al.* 1990).

Critically, the focus on wildlife, and hence on the very few communities inside wildlife management areas (van der Heiden 1991), has left the majority of rural, natural resources-dependent communities outside of the CBNRM programme (Cassidy 2000). This exclusion from the focus of the country's CBNRM approach, combined with the disempowerment of traditional local level leaders through the centralisation of governance since independence in 1966, and the difficulty of implementing national policies in remote areas, has resulted in unmanaged, open access to natural resources in the majority of rural settlements.

By engaging community members in direct monitoring and management based on knowledge they already have (Phuthego and Chanda 2004) provides a unique opportunity to avoid the errors introduced by formal CBNRM programmes; they can be made aware collectively of how their activities affect the environment (Folke *et al.* 2002; Plummer and Armitage 2007). Further, by starting with the assumption that systems are dynamic, and that resource availability varies, those in rural communities can be given the skills to balance environmental change with human needs (Olsson *et al.* 2004). However, in common property systems, where incentives to invest management efforts may be undermined by others maximising their off-take (Ostrom *et al.* 1999), focusing on the political institutions that condition people's behaviour and control their rights is perhaps more important (Peterson 2000; Adhikari 2005). Gender, class, ethnicity, resource interest, and existing power relations are some of the many factors that influence how decisions relating to resources management are made. If all people are to support resources management, they need to feel that their likely benefits from such management are equitable (Goodland and Daly 1996). Underpinning this sense of equity, is the need for good governance, based on participation at all levels (Leach *et al.* 1999; Levin 2006). Recent research in the region supports the notion that good governance should be based on transparency, accountability, and democratic participation in decision-making (Child and Barnes 2010; Collomb *et al.* 2010). Here we document the process of building sound governance practices in a fledgling community organisation tasked with managing the natural resources in its area.

Study site

The study is based in Tubu Village in the Northwest District of Botswana. The village lies to the west of the Okavango Delta (Figure 4.1), on communal tribal grazing land. It is a small community of about seasonal flood plains. Most of the people are of Yei origin, and are primarily crop farmers but also keep livestock, fish in the river and lagoons, and harvest river and veld resources. Due to the extreme remoteness of the settlement, reliance on the natural resources base is still very high.

Tubu is separated from the adjacent wildlife management area (WMA) by a veterinary disease control fence. The WMA land just inside the fence is now a private commercial tourism lease area, where a corporation conducts safaris for profit. Although the fence was installed prior to the designation of the WMA and the awarding of the lease, the interplay of truncated access due to the fence and the new land rights created conflict between the community and the safari operator. In order to explore collaboration between the two parties, the Tubu Joint Management Committee (JMC) was formed in 2008 by a University of Botswana GEF-funded project (BIOKAVANGO), which aimed to build local capacity for

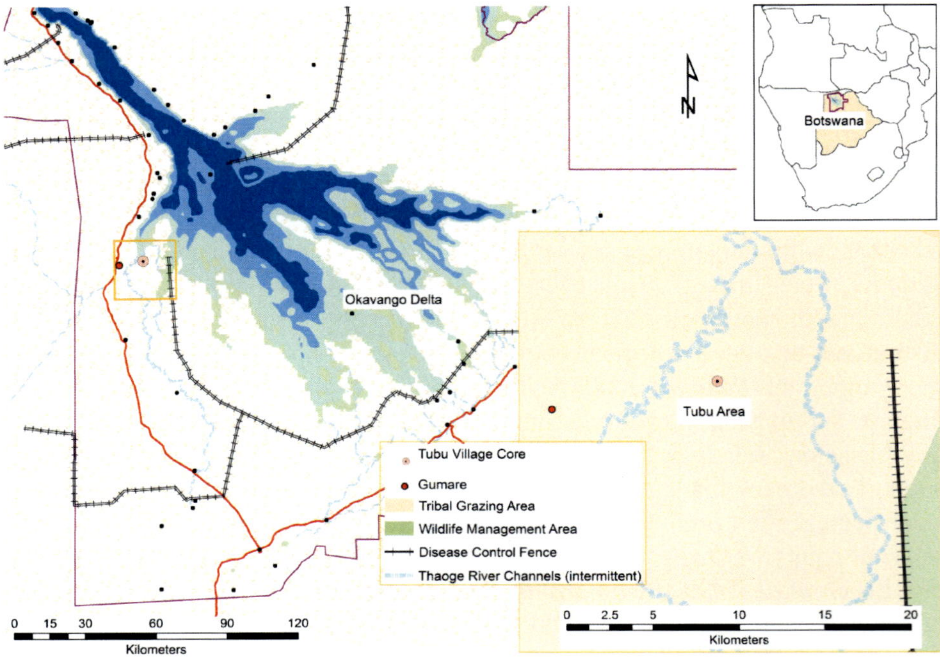

Figure 4.1: Map of the Okavango Delta showing Tubu Village in the west, some distance from the main tarred road, and located between two distributaries of the seasonally flooding Thaoge River. Source: Prepared by Dr. Lin Cassidy

natural resources management. The committee was to be used as a good governance instrument fostering communication, transparency, accountability participation, trust building and resolving conflict; all of which are principles of good governance. The JMC had members from the village leadership, ordinary villagers formed into a village trust and other relevant service providers and resource users (see Table 1). However, during this process, it emerged that what was more important for the Tubu community was a better understanding of, and managed access to, the resources immediately surrounding the village – on their side of the fence. The following objectives were therefore set for the JMC:

- To manage and protect the natural resources of the Tubu area;
- To facilitate the provision of bulk infrastructure and social services to Tubu Village;
- To facilitate the creation of new livelihood opportunities through tourism;
- To facilitate the improvement of agriculture (livestock and cropping); and
- To ensure good governance.

The committee comprised all stakeholders of land resources management of the area in order to open up communication and resolve conflict over land resources. Its current activities include fundraising for operating resources; mobilising com-

Table 4.1: Members of the Tubu Joint Management Committee (JMC as initiated in 2008.

	Institution/Department	Name of the nominated JMC Member	Position
1	Department of Forestry and Range Resources (DFRR)	Mr. Mafoko	Range Ecologist, Maun
2	Department of Tourism (Maun)	Ms. L. Ntshekisang	District Tourism Officer, Maun
3	Department of Wildlife and National Parks (DWNP)	Mr. Masunga	Wildlife Biologist, Maun
4	District Commissioner (Gumare District Administration)	Mr. P. Makambe	District Officer, Gumare
5	Botswana Police (Gumare)	Mr. Hulela	Inspector, Gumare
6	District Agricultural Office	Mr. Tjitimesa	Tubu Agricultural Extension Officer, Gumare
7	Tawana Land Board	Mr. Matlhogonolo Moshagane	Land surveyor, Gumare
8	KgosiyaTubu (Tubu Headman)	Kgosi Motshidiemang	Tubu Headman
9	NG 25 Concessionaire	Mr. D. Kays	Maun resident
10	Tubu Councillor	Mr. M. Karapo	Tubu Councilor
11	Tubu Village Development Committee (VDC)	Mr. G. Xaa	Tubu resident
12	Tubu Village Extension Team (VET)	Mr. C. P. Molefi	Tubu resident
13	TubuFarmers Committee	Mr. M. Masarozi	Tubu resident
14	Tubu Community Conservation Committee	Mr. B. Merapelo	Tubu resident
15	*Tubu Youth Committee	Ms Esther Setshego	Tubu resident
16	Biokavango Project	Staff Member	Maun Resident
17	Trust Committee	Ms. O. Boitshwarelo	Tubu resident
18	Trust Committee	Ms. K. Sashone	Tubu resident
19	Trust Committee	M. Boikwathao	Tubu resident
21	Trust Committee	Mr. O. Boitshwarelo	Tubu resident
22	Trust Committee	Mr. O. Mabe	Tubu resident
23	Tubu Farmers‘ Committee	Mr. M. Mosarozi	Tubu resident

munity planning to formulate plans and strategies to achieve the stated objectives; and monitoring community resources through the management oriented management system (MOMS).

Good natural resources governance

According to Dietz *et al.*, better governance of the commons can be achieved when "the resources and use of the resources by humans can be monitored", and when communities "maintain frequent face-to-face communication and dense social networks … that increase the potential for trust…" (2003: 1908). In order to establish trust, it is important to ensure transparency and accountability, and this work builds on the hypothesis that active participation in the management process – participatory democracy – is more likely to succeed than representational governance. This implies that not only the benefits, but also the responsibility for sound management, must be shifted to the local level (Child and Barnes 2010).

Communication and trust can be understood by assessing the involvement of community members in decision-making, and by their knowledge of, and attitudes towards, the representative organisation, particularly in terms of transparency and accountability. This study uses the two-stage "Dashboard" approach to investigate the extent to which the fledging JMC practices good natural resources management (participatory democracy) and promotes awareness of and participation in Tubu's natural resources management activities. It also sets a baseline for future assessments of community involvement, so that changes in transparency and accountability may be noted and appropriate action taken.

As stated above, the Dashboard approach comprises a two-stage process. The first is a detailed household survey on awareness of, participation in, and attitudes towards the activities of the community organisation responsible for natural resources management. The second is the presentation of these results at a full community meeting where community members are shown the results, and asked to tell the organisation's representatives whether or not they are satisfied with these.

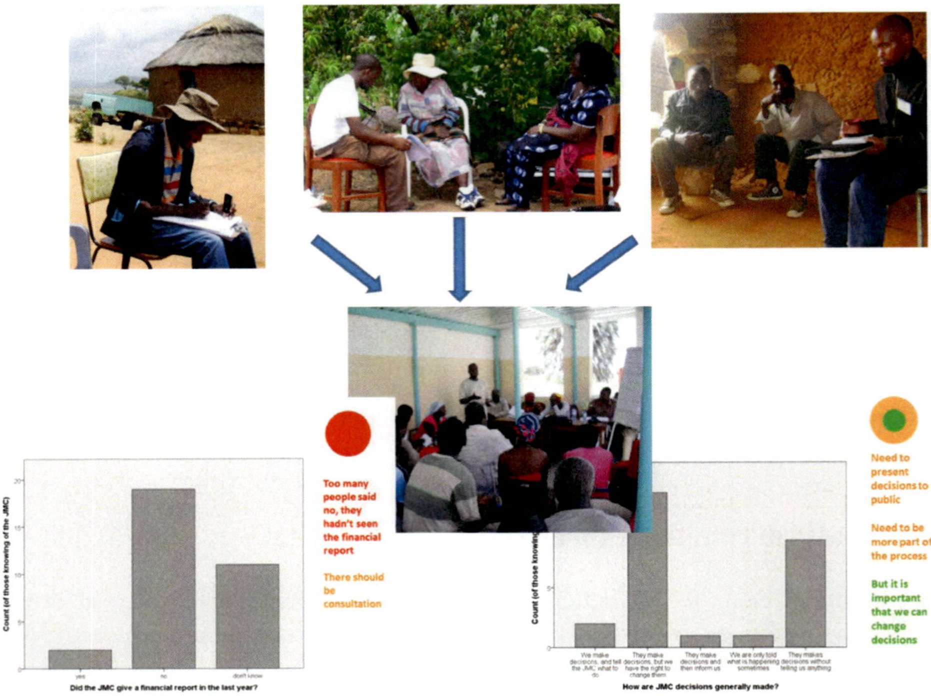

Figure 4.2: A visual representation of the Dashboard approach – A household survey followed by a community meeting presentation of results.

Community governance in the context of the DARMA project

The scientific issue that DARMA addresses is the fragmentation of the knowledge base and, hence, management policy in Southern African commons. Traditionally management has been carried out by government agencies that focus on a particular sector, such as fisheries or forestry, using knowledge from scientists who specialise in that sector, often without working with community groups to create and enforce rules developed for the sector. With the advent of the ecosystem approach to NRM, which is based on a growing appreciation of the interrelatedness of ecological, social and economic dynamics of natural resources, new cross-disciplinary scientific networks are needed that can identify specific areas of interconnections between sectors, articulate and implement the research required, and translate the results into advice for policy makers

The overall broader objectives to which DARMA contributes are the protection of biodiversity and poverty alleviation on exploited commons. Integral to achieving these broad objectives is conflict resolution, because ecological degradation leads to competition for the remaining resources. Therefore, effective integrated commons management addresses three related sets of problems simultaneously: a) limiting access to resources to prevent overexploitation; b) economic development that relieves poverty while respecting those limits; and c) managing conflicts that arise when access to resources is limited. All three of these problems require knowledge. The limits must be set through the work of natural scientists, both poverty alleviation and the effective implementation of management require the work of social scientists, and managing conflict requires involving community stakeholders or resource users who know locally appropriate solutions.

Implementation of DARMA in the Okavango Delta sought to establish a baseline for good natural resources governance for a relatively new community-level resource management institution. This was done by setting standards for good governance by which the community can periodically evaluate its own institution. The expectation is that involving stakeholders in setting, monitoring and evaluating good governance thereby reduces scope for conflict over management and access to resources.

Methods

In late October 2011 a survey was conducted of all 125 households in Tubu. Survey questions focused on basic demographic and economic characteristics, and general levels of community involvement, before moving on to questions about opinions and awareness relating to governance, which served as indicators of participation, transparency and accountability of the JMC. The questions focused on the correct identification of officers, satisfaction with meetings, correct identification of rights and responsibilities, and satisfaction with financial management, among others. The survey also collected information on perceptions of animal population trends and interactions, for cross-checking against crop and livestock depredation and poaching incidents recorded elsewhere.

The questionnaire contained a section with questions about more general awareness that did not only reference the JMC. However, much of the information on environmental aspects is collected through the JMC. There are two aspects that affect information dissemination. First, the traditional approach to information dissemination is through the *kgotla*, and we have already seen that attendance is influenced largely by whether the household is located in or outside of the village core. Second, there is the issue of whether information presented is viewed as important or not. It is clear from the results that in a dispersed village like Tubu, the *kgotla* system may not be working in terms of information dissemination, and that other methods should be considered for improved communication and transparency, such as sending pamphlets home with school children.

In April 2012, results of the survey were taken back to the community, and presented at a formal meeting, with the JMC representatives present. As each indicator was presented, community members were asked to discuss whether they think the result was good, neutral, or bad, and how they would like the JMC to do things differently. This process allows ordinary community members to hold their representatives publicly accountable, so that expected changes are more likely to be implemented. It is at this stage that the analogy of a dashboard is introduced. In order to make the results visual, and to accommodate the varying levels of literacy found in rural communities, community members are asked to think of running the organisation as similar to driving a car – all passengers (community members), and not only the driver (community leader), are responsible for being alert to possible danger such as a donkey straying onto the road or the smell of a hot engine. The dashboard symbolises the need to continuously monitor the state of the vehicle (community organisation). If a red icon lights up on the dashboard, this signifies that something is wrong, and that immediate action is required. An orange light is a warning that corrective measures should be put in place. A green light means that all is well, and the journey can proceed.

Results

Household survey

Demographic characteristics

Tubu village comprises 125 households, two-thirds of which are located within the core of the village, and one-third some distance outside land areas and cattle posts. 62% of households are male-headed and 38% female-headed indicating a fairly traditional society where marriage is still an important determinant of the independence of a household. It could also indicate a below national average level of male migration. Nearly all (93.6%) households belong to the Yei ethnic group. On average, households comprise between five and six members, of whom three are adults. The mean dependency ratio is low at 0.93 children for every adult. However, there are strongly significant differences ($F = 9.851$, $p = .002$) between male- and female-headed households; the former have an even lower mean dependency ratio at 0.72, while the latter have on average 1.3 children for every adult. This is not immediately understandable; however it may be because females tend to take in grandchildren more often.

Formal education varies strongly within the community. 58% of household heads have no formal education. This is strongly linked to age ($F = 9.079$, $p < .000$), with older household heads less likely to have formal education. However, 74% of households have at least one member who has completed primary school. Only 26% of households had a member in formal wage employment.

Most houses are roofed with thatching grass (82%) while 18% have corrugated iron. 48% of houses are built with mud bricks made locally, while 36% are built with reeds. Only 16% are built with cement blocks or bricks. Almost a quarter of all households had a water standpipe in their own yard. However, most use communal standpipes, and 19% collect water from the nearby river. The majority of households (62%) have no toilet facilities and use the surrounding bush.

92% of households usually grow crops. Field sizes are quite skewed, with most people having small fields, and a few having really extensive lands. The median field size owned is 1.7 hectares. Maize is the dominant cereal crop, and only 6% of respondents planted sorghum in the preceding season. The mean reported yield for maize was 680 kg/ha. 82% of households keep livestock. Ownership of cattle is skewed, with 50% of households owning less than 10 cattle each, and only 9% of households owning more than 50 cattle each.

Awareness of the JMC and its code of conduct

Only 26% of households had heard of the JMC. This was strongly dependent on the location of the household, with those living outside the village core far less likely to know of it ($Q^2 = 8.041$, $p = .005$), and far fewer outside the village core able to name the JMC chair correctly ($Q^2 = 7.467$, $p = .006$). Location within or outside the village core appears to be a critical factor influencing awareness – non-JMC participation shows similar results: frequency of church attendance and frequency of *kgotla* meeting attendance was significantly higher for households within the village ($F = 5.512$, $p = .021$ and $F = 5.866$, $p = .017$ respectively).

Female respondents were also less likely to have heard of the JMC ($Q^2 = 3.348$, $p = .067$). This is interesting because there were no significant differences between male and female respondents and their awareness of the VDC and its members. On the other hand it may be due to the fact that land resources management in Botswana is male dominated. Even the JMC had 13 Tubu-based members of whom 10 were male at the time of the study. It must be noted, though, that there were no significant differences between gender of household head and re-

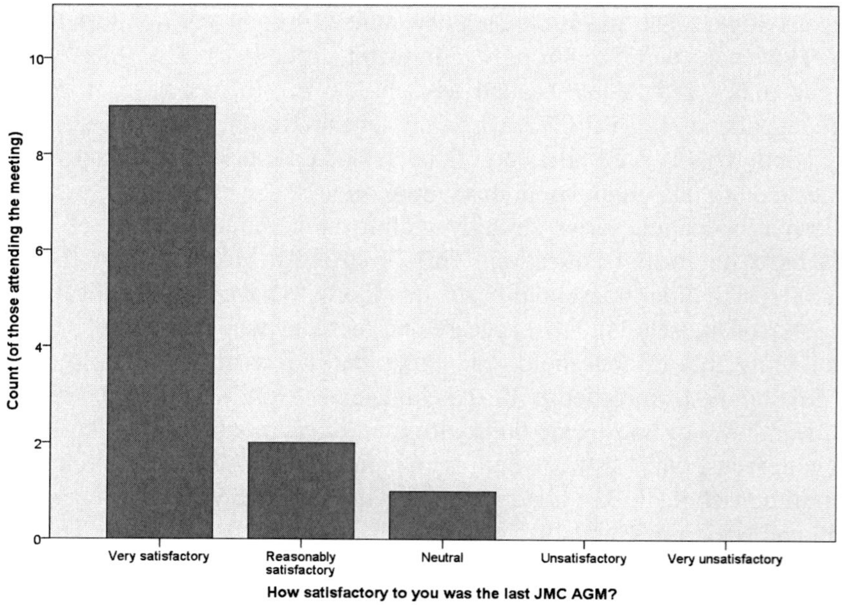

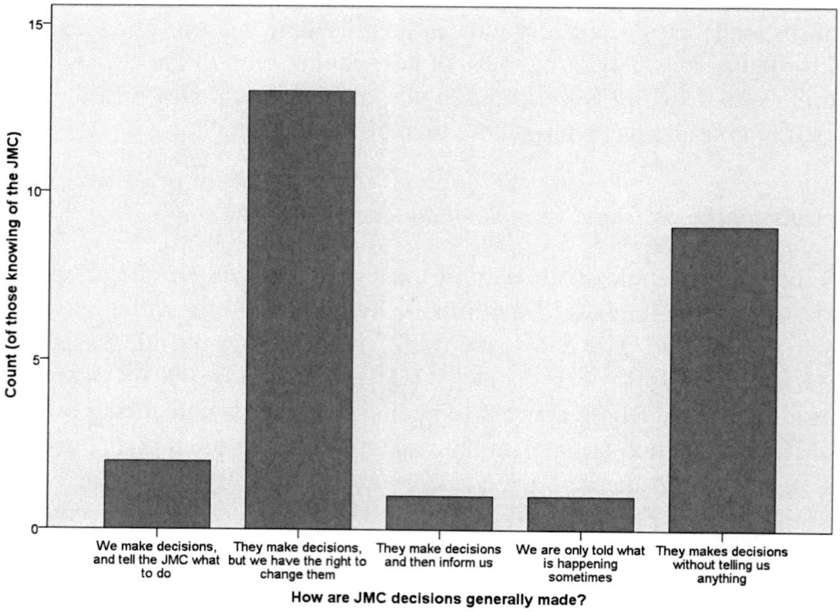

Figure 4.3: Graphs showing how participating householders feel about their involvement in the JMC. Note: Figure 4.3 a) shows that most people were pleased with the last annual general meeting, while Figure 4.3 b) suggests some divisions between those who feel some influence versus those who already feel marginalised.

spondents' awareness of the JMC. This means that at least some members of all households appear to have some knowledge, but that women appear to have less.

Of the 26% who know of the JMC, 38% had attended the last annual general meeting (AGM) of the JMC. Most (88%) were able to name the JMC chair correctly. 57% said they were consulted during the development of the JMC code of conduct, and of these, 64% (14 households) were satisfied with it. This shows that communication is a useful tool to build trust.

Decision-making

For those already participating in the JMC, opinions are generally favourable towards it (Figure 4.3). People appear to be happy with meetings, and feel they have some influence over decision-making. However, the responses were split almost equally when evaluating the quality of the decisions made by the board, with half saying they were good, and half saying they were selfish.

Accountability

Of those respondents who knew about the JMC, most were not sure about the overall running of the JMC by its representatives (Figure 4.4). Of concern is that while some respondents were aware that a financial report had yet to be made to the community, many did not know either way.

Community participation

On average, most householders attend *kgotla* (formal community) meetings about seven times a year. Some 25% do not attend these meetings at all. Distance is clearly a factor in this: householders from beyond the village core attended meetings significantly less frequently ($F = 5.866$, $p = .017$). Most householders are members of a church, but there are nine different church groups in the community, so this is not a central networking function. About 57% of households have a member in at least one of the community institutions, such as the VDC, health committee or crime prevention committee.

Community member rights

In terms of rights and responsibilities, of those who knew of the JMC:
- 77% knew they had the right to stand for election
- 78% felt they had the responsibility to hold committee accountable
- 84% felt they had the responsibility to actively participate in meetings
- 78% felt they had the duty to report illegal plant or animal use.

However, 48% did not believe they had the right to decide on how to use JMC money, and 45% believed they did not have the right to check on how JMC money was spent. These are significant proportions of potential JMC members who did not believe in financial accountability.

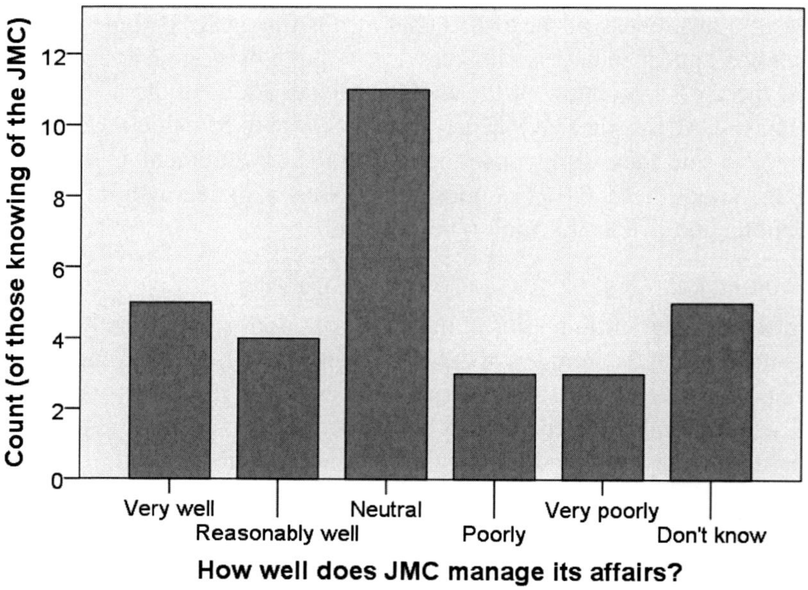

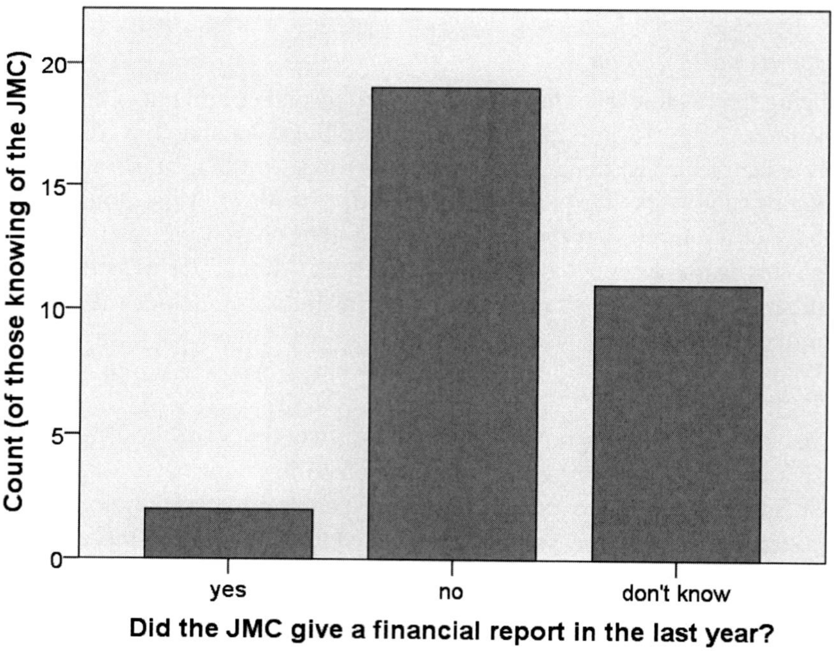

Figure 4.4: Graphs showing the perceptions of households who know of the JMC, with regard to a) the general running of the organisation's activities, and b) their knowledge of finances – no report had yet been produced.

Financial management

Amongst those that were aware of the JMC, there was generally a positive feeling towards the JMC programme (Figure 4.5). However, when looking specifically at financial management, it is clear that, firstly people are mainly unaware of how financial issues are being handled, and this is likely the reason why in most cases where people had an opinion, it was negative (Figure 4.5). In part this is due to the early stages of the programme, as is also reflected in people's attitudes towards benefits arising from the JMC activities (Figure 4.6). As stated earlier these results will set the baseline for future monitoring and evaluation.

In terms of how well the finances were presented, 37.5% of those knowing of the JMC claimed (correctly) that the finances had not been presented at all, while a further 53.1% did not know. Linked to this is the fact that 78% of respondents did not know how well the JMC finances are accounted for – however, the remaining 22% were fairly well split, with 9.4% saying finances were well accounted for, 9.4% saying finances were poorly accounted for and 3.2% were neutral on the issue. Only one respondent knew how much money was spent on JMC administration in the past year. Again it must be reiterated that this is largely because the JMC has no income of its own, and hardly any turnover, with administrative costs having been carried largely by outside organisations (donors).

Comparison of perceptions of the JMC with perceptions of other community organisations

While 57.6% and 45.6% of respondents claimed to know the names of the VDC chair and secretary respectively, only 12.7% and 7.9% said they knew the names of the JMC chair and secretary. To a large extent this is a reflection of the current levels of functionality of the two organisations. Another indication of the limited activities of the JMC is that, while about 90% of those who said they know the names for the VDC chair, VDC secretary and JMC chair were correct, only 31.3% of those who said they knew the JMC secretary were able to name him correctly.

When comparing the JMC to the Land Board (an external institution) and the VDC (internal) in terms of how management of affairs are perceived, the Land Board has a more favourable image, while the JMC is comparable to that of the other internal institution, the VDC; in fact the JMC was regarded by a slightly higher proportion of respondents as being well managed. This could have to do with the authority and resources associated with the institution. While the JMC is a fledgling organisation, its members (see Table 4.1) include individuals from higher authority and more resourced organisations. Hence the JMC may be viewed as having the potential to be more organised than the rather resource-starved local organisation, the VDC. Moreover the JMC had backing.

Perceptions on decision-making vary in a similar manner: most people (70%) feel Land Board decisions are good, while only a minority feel the same way about the VDC and JMC decisions. It is noteworthy that more people are unsure

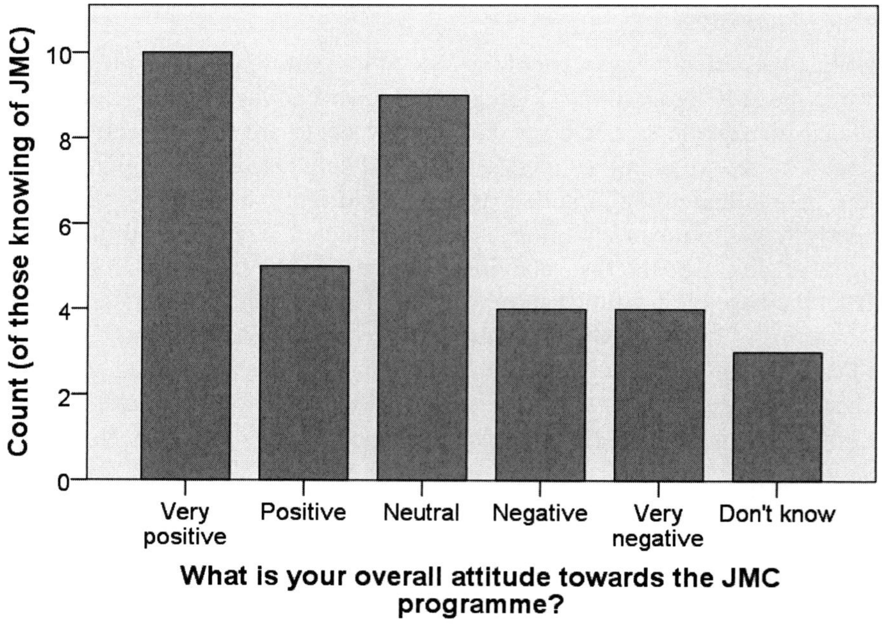

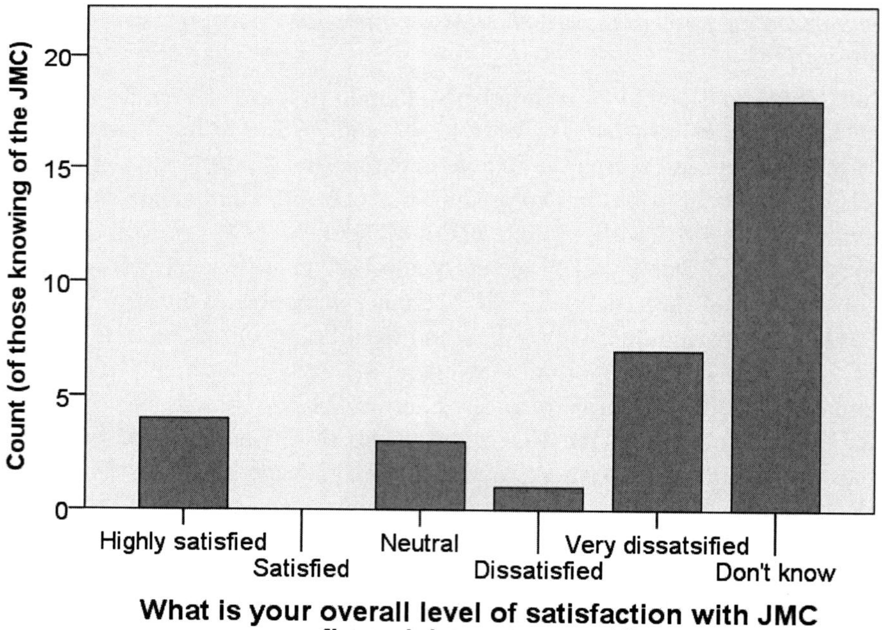

Figure 4.5: Graphs showing the overall attitudes of those Tubu householders who were aware of the JMC towards a) the overall programme, and b) towards financial management.

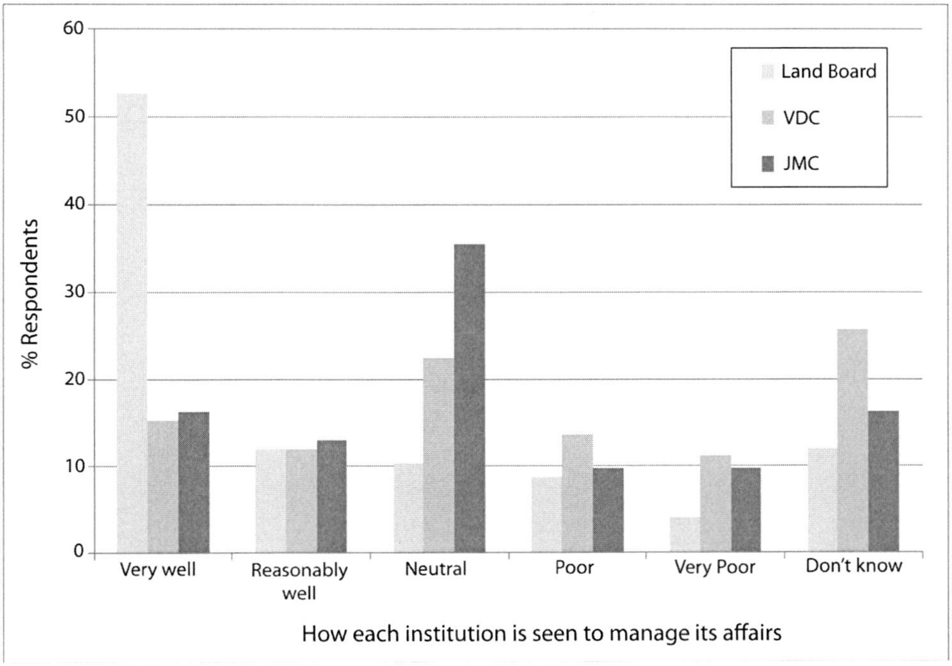

Figure 4.6: Comparison of the Land Board, Village Development Committee and TubuJoint Management Committee in terms of how well they are perceived to manage their affairs.

about VDC decisions than they are about the decisions of the JMC. These results are also not immediately understandable as usually people of Ngamiland are unhappy with the decision of the Land Board. This could be due to a village dispute involving the VDC. Another study (unpublished) conducted by the University of Botswana in 2010 in the same village found the VDC was struggling to make the top five most useful local institutions. This is unusual as the VDC and the *Kgotla* (traditional meeting place and seat of the Chief) often make the top three local institutions in such remote settlements. However VDCs sometimes fall into controversy when they are highly politicised and controlled along partisan lines or when certain members are perceived to have 'overstayed' by a section of the community. There have also been other perceptions and accusations that VDCs misuse their positions and are guilty of favouritism when recommending community members for government aid or employment and that they misuse community resources, especially finances. These factors could cause the VDC to fall out of favour with the community as appears to have happened here.

Information dissemination from environmental monitoring activities

Table 4.2 shows that there are significant differences between the village core and periphery in terms of receiving environmental information. However, it is

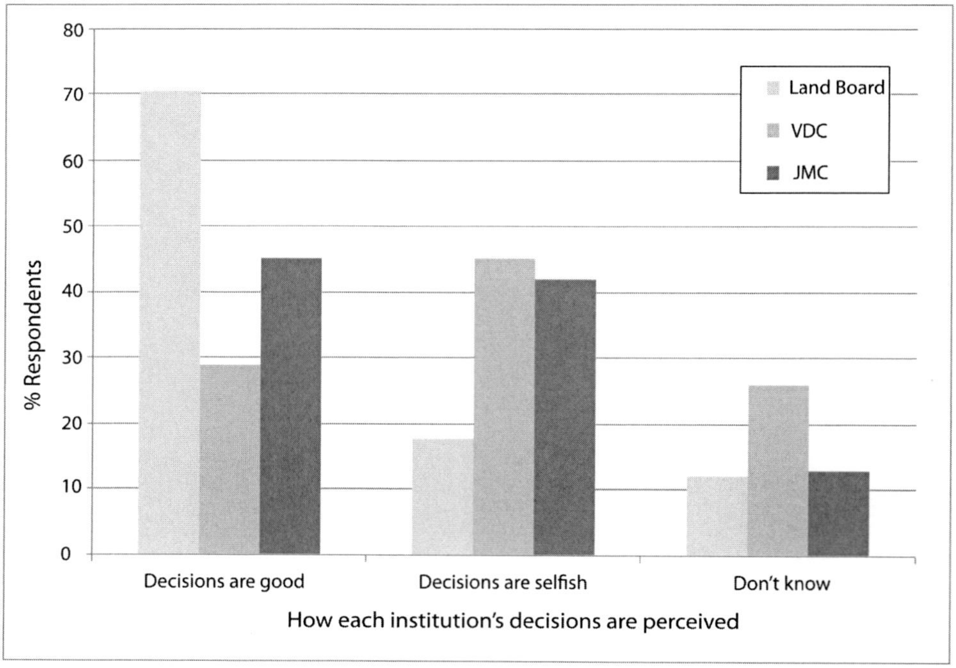

Figure 4.7: Comparison of people's perceptions of decision-making by the Land Board, Village Development Committee and Tubu Joint Management Committee.

Table 4.2: Differences between householders in and outside of the village core in terms of accessing the information they need. Significant differences are given in bold.

Topic of information	% in village core getting all or some information	% outside village core getting all or some information	X^2	*p*
Problem animals	54.0	28.6	6.273	.043
Trends in animal populations	40.5	17.6	5.494	.064
Trends in endangered animals	51.3	17.1	12.078	.002
Trends in livestock mortalities	53.3	23.5	10.810	.004
How MOMS rangers are working	36.2	9.4	8.464	.015
Trends in poaching	64.5	25.7	14.756	.001
Use of important plant species	64.5	26.5	13.803	.001

clear that where people are able to access the *kgotla*, some key information is getting through to at least half of the households. Notably, awareness of the MOMS rangers, who were engaged at the same time as the formation of the JMC to collect monitoring data (Magole *et al.*, this volume) is quite low. This corresponds with the low visibility of the JMC as reported earlier. It could also have to do with

Maize is an important crop. Maize yields are higher in flood plains under the traditional flood plain cultivation system. Image: Alyne Delaney

the fact that while the survey took place in the village the rangers operate at the resource use areas which are largely on the outskirts.

Perceptions of land use zones and wildlife

Of those who knew of the JMC, 69% said they had been consulted during the development of the management plan, and 84% understood that it had different land use zones for wildlife and farming activities. Most people were satisfied with the size of the area set aside for wildlife-based activities, but 33% felt that the area was too big and should be decreased.

Zoning is relevant to the community's management of human-wildlife conflict. In the previous year, 32% of households experienced some depredation on their livestock, while 30% experienced depredation on their crops. Two households reported a family member injured by wildlife in the previous 12 months.

People's perceptions of population trends of different species appears to be somewhat influenced by whether those animals are problematic (i.e., predating on livestock or crops) or not. However, some perceptions matched observed trends. Importantly, 97% of people felt that the number of elephants was increasing. But those living within the village core were more likely to feel that the number of lion were increasing than those living outside (Q^2= 6.157, p = .046) – yet most predation is likely to occur further from the village core. Differences between core and periphery responses to awareness of wildlife population trends are given in Table 4.3 below.

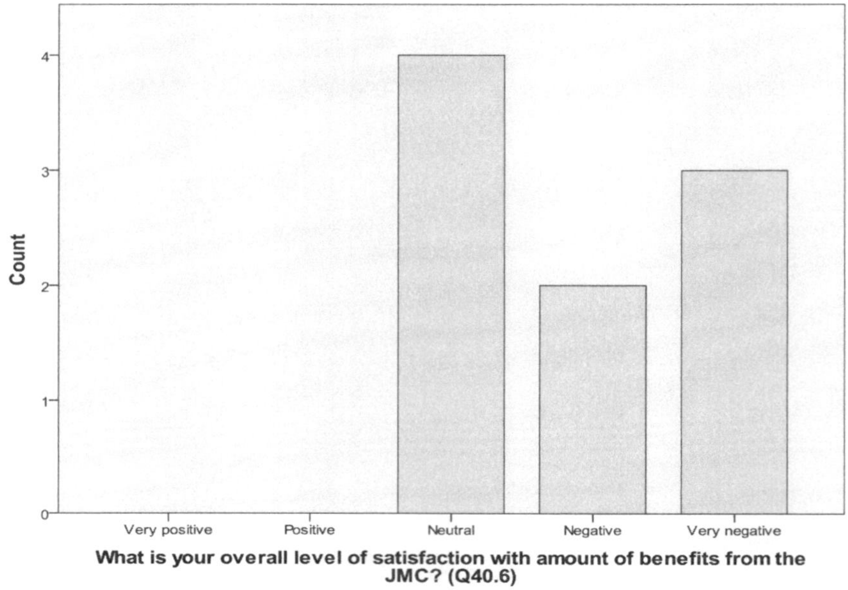

Figure 4.8: Levels of satisfaction with benefits from the JMC

Table 4.3: Differences in of wildlife population trends between households within and outside of the village core. Significant differences are given in bold.

Species	% in village core seeing an increase	% outside village core seeing an increase	X^2	*p*
Elephants	97.4	97.6	.738	.691
Lion	83.8	65.8	6.157	.046
Hyena	78.7	61	4.290	.117
Hippo	85.5	92.5	2.180	.336
Baboon	74.3	76.9	1.887	.389
Porcupine	81.9	90	3.004	.223
Wattled crane (endangered, not a problem species)	41.1	45.5	4.202	.122

Overall satisfaction with the JMC

Interpretation of findings on overall satisfaction with the JMC must bear in mind that only a small proportion of people (n = 31) actually knew about the JMC in the first place. However, it is clear that people's expectations are not being met, as no respondents expressed overall satisfaction with the JMC, showing the importance of managing participation and transparency.

Community feedback

The community feedback took place at the village *kgotla* (meeting place), with the JMC committee members and village headman and ward headmen present. Approximately 50 community members attended to help evaluate the survey findings. Two facilitators presented hand-drawn graphs on flipcharts of the summary findings presented above. While one facilitator elicited the evaluation, the other marked the charts with the colour the community called out, and noted comments.

6.2.1 Evaluation of survey results

The outputs of the community dashboard evaluation are given in Figure 4.9 below. The community felt that it was bad (amber with red edges) that more people had not attended the JMC, and explained why: there were a lot of mistrust, but they also mitigated this statement by saying that the timing had been poor. They were also concerned (a little amber surrounded by lots of red) that so few people knew of the JMC, but understood that this was a process of education and raising awareness. They were pleased (green) to note that most of the participants at the last JMC AGM had been satisfied with the meeting. They felt it was a source of concern (green surrounded by amber) that decisions were made by the committee and not by all members, but felt it important to point out the positive, that sufficient people knew they had the power to change decisions. They noted that it was early days, and that there was limited money, so expressed only slight concern (red with amber edges) over the fact that most felt the JMC affairs were not well managed – but someone pointed out that a further cause of concern was that too many people had responded "don't know", suggesting a reluctance to express unhappiness.

One of the most interesting discussions at the feedback meeting revolved around the presentation of a financial report – or in this case, the fact that one had not been presented. This was labelled strongly as a critical or stop situation (red), and a call for change. When the committee representative tried to justify the lack of a financial report by saying it was because there was virtually no funding passing through the JMC (a verified fact), people asked how they were to know this if they did not see the report. The facilitated exchange was a really important learning experience for both community members and the JMC committee, who suddenly realised the importance of transparency – that the amount of the finances was not the issue, but rather that people needed to know the state of the finances.

This point came up again when looking at satisfaction with financial management, and the community members felt that too many had said "don't know", and that even if there wasn't much to report, the community must be reported to.

People were generally satisfied (green) with the responses to knowing their rights with regard to decision-making. Their positive evaluation was based in part on awareness levels, but also on the rights themselves, which they felt were sound. They also gave a critical alert to the fact that most had not seen benefits, and that indeed, there were no tangible benefits on the ground yet. They felt that people would need to start feeling the incentives in order to give continued support to the JMC.

JMC committee acceptance of the dashboard evaluation

As this was the first time that a dashboard evaluation was conducted in Tubu, it was clearly a learning process for both general members and the committee. As could be expected, the initial response for the committee was to feel defensive, and to try and justify the negative responses. However, with interventions from the facilitation team and most importantly, from the village Headman, the committee was able to see that the feedback was not intended as a criticism, but to provide a path for democratic participation. Critically, the Headman chose to return to using the analogy of a vehicle being steered, but with the passengers having both the right and the responsibility to ensure that the vehicle was driven properly. The committee was able to see that they now had some guidance on where to target their efforts to be more representative of their constituents, and what activities they needed to follow to ensure the buy-in of the broader community base.

Discussion and conclusion

The assessment of the JMC's governance at this stage is primarily intended as a description of baseline conditions against which progress and change can be monitored. However, there are still important lessons that can be learned from current attitudes and perceptions.

At this stage the most important aspect of participation is raising basic awareness of the JMC's existence. Where people are included, they predominantly feel positive towards the JMC and its functioning. However, given that about three-quarters of respondents claimed not to have heard of the organisation, there is clearly a lot of basic mobilisation required. Of particular concern is ensuring the involvement of outlying households. Since Tubu is a very scattered settlement, it is not easy to disseminate information, and yet this is vital for visibility, legitimacy and obtaining support for facilitating managed natural resources access.

The overall positive attitude toward the JMC programme reflects the value that people ascribe to natural resources, and their proper management. This positive response is particularly encouraging given the conflict situation that had led to

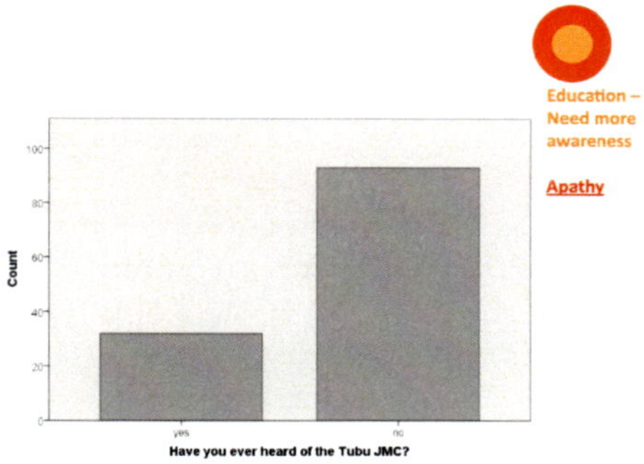
Education –
Need more
awareness
Apathy
Count
yes
no
Have you ever heard of the Tubu JMC?

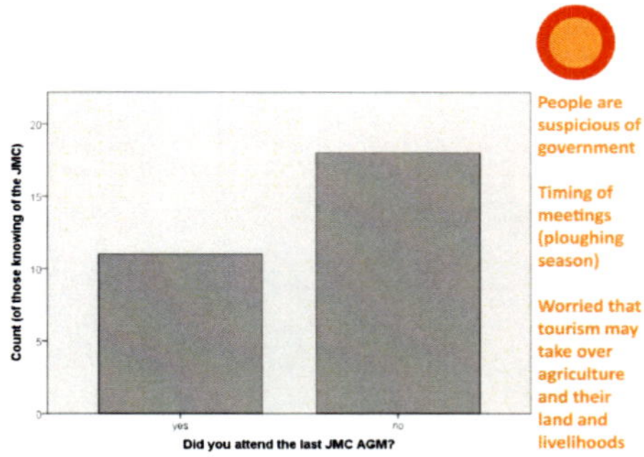
Count (of those knowing of the JMC)
yes
no
Did you attend the last JMC AGM?
People are suspicious of government
Timing of meetings (ploughing season)
Worried that tourism may take over agriculture and their land and livelihoods

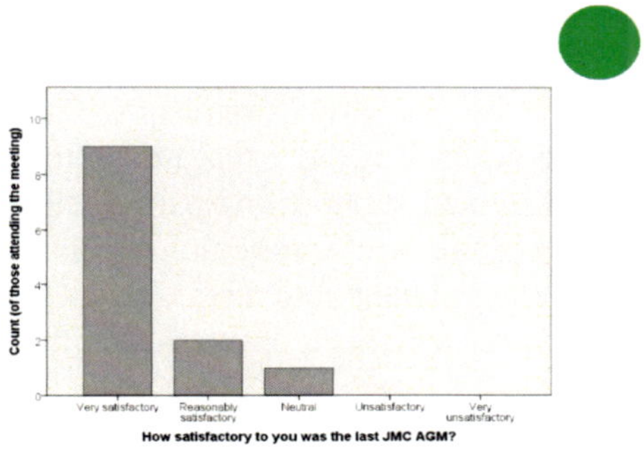
Count (of those attending the meeting)
Very satisfactory
Reasonably satisfactory
Neutral
Unsatisfactory
Very unsatisfactory
How satisfactory to you was the last JMC AGM?

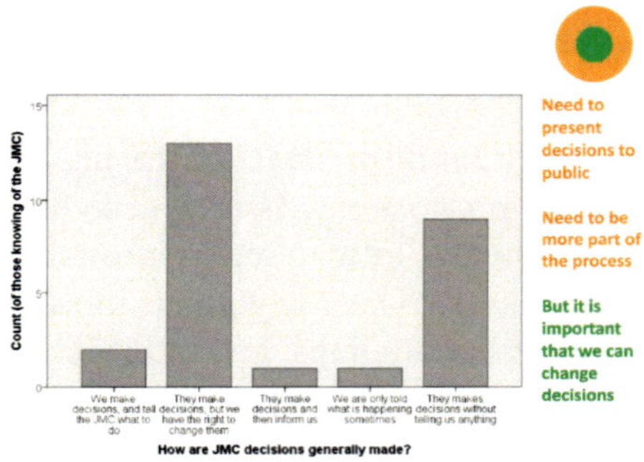
Count (of those knowing of the JMC)
How are JMC decisions generally made?
Need to present decisions to public
Need to be more part of the process
But it is important that we can change decisions

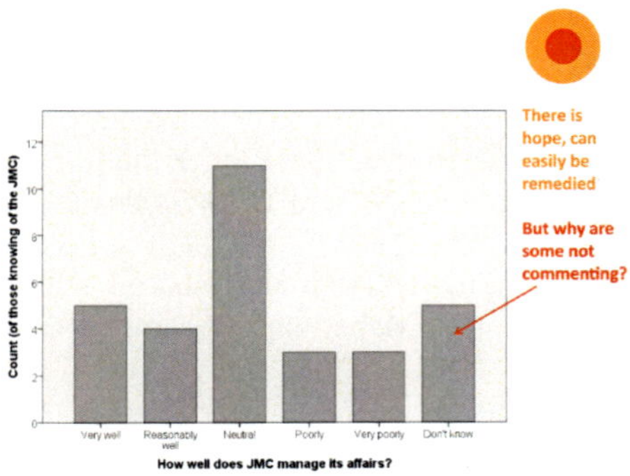
There is hope, can easily be remedied
But why are some not commenting?
Count (of those knowing of the JMC)
Very well
Reasonably well
Neutral
Poorly
Very poorly
Don't know
How well does JMC manage its affairs?

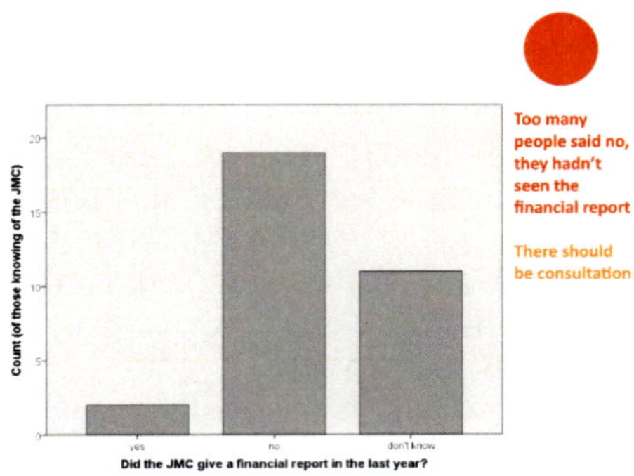
Count (of those knowing of the JMC)
yes
no
don't know
Did the JMC give a financial report in the last year?
Too many people said no, they hadn't seen the financial report
There should be consultation

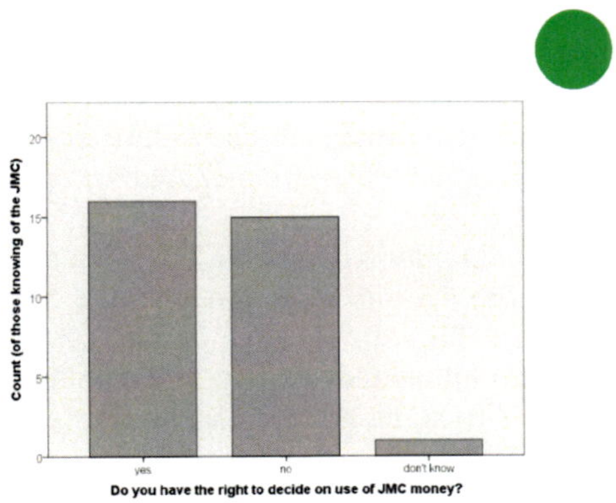
Count (of those knowing of the JMC)
yes
no
don't know
Do you have the right to decide on use of JMC money?

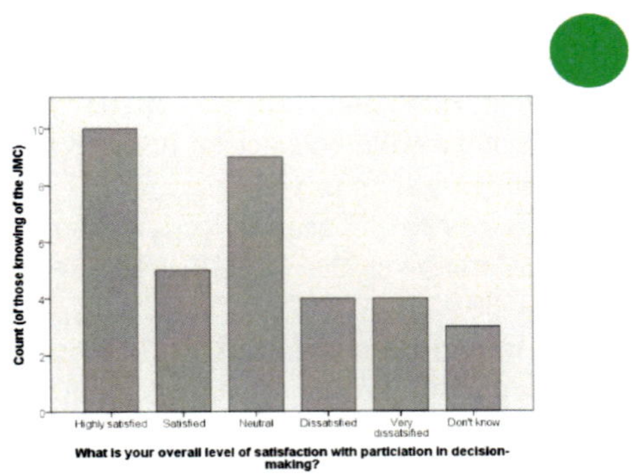
Count (of those knowing of the JMC)
Highly satisfied
Satisfied
Neutral
Dissatisfied
Very dissatisfied
Don't know
What is your overall level of satisfaction with particiation in decision-making?

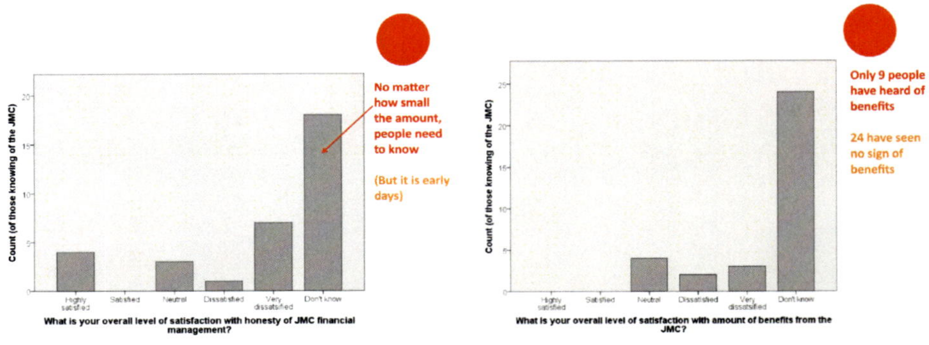

Figure 4.9: Community evaluations of the results of the survey

the formation of the JMC in the first place. However, it is clear that maintaining financial transparency is critical to ensuring sustained support and participation. Even if there is little to report, providing details of financial transactions will help dispel distrust. These key points must be made clear at the public presentation and assessment of findings, and the JMC representatives will need to internalise them into their management activities.

References

Adhikari, B. (2005). "Poverty, property rights and collective action: Understanding the distributive aspects of common property resource management", *Environment and Development Economics* 10: 7–31.

Balint, P. J. and Mashinya, J. (2006). "The decline of a model community-based conservation project: Governance, capacity, and devolution in Mahenye, Zimbabwe", *Geoforum* 37(5): 805–815.

Blaikie, P. (2006). "Is small really beautiful? Community-based natural resource management in Malawi and Botswana", *World Development* 34(11): 1942–1957.

Cassidy, L.(2000). Community Based Natural Resources Management and Legal Rights to Resources in Botswana.Gaborone, SNV/IUCN CBNRM Support Programme. Occasional Paper No. 4.

Chabwela, H. N. and Haller, T.(2010). "Governance issues, potentials and failures of participatory collective action in the Kafue Flats, Zambia", *International Journal of the Commons* 4(2): 621–642.

Child, B. (1996). "The practice and principles of community-based wildlife management in Zimbabwe: the CAMPFIRE programme", *Biodiversity and Conservation* 5(3): 369–398.

Child, B. and Barnes. G.(2010). "The conceptual evolution and practice of community-based natural resource management in southern Africa: past, present and future", *Environmental Conservation* 37(3): 283–295.

Child, B. and Peterson, J.H. (1991). CAMPFIRE in rural development: the Beitbridge experience. Harare, Zimbabwe, Department of National Parks and Wildlife Management.

Collomb, J. G., Mupeta, P., Barnes, G., and Child, B. (2010). "Integrating governance and socioeconomic indicators to assess the performance of community-based natural resources management in Caprivi (Namibia)", *Environmental Conservation* 37(3).
Dietz, T., Ostrom, E., and Stern, P.C. (2003). "The struggle to govern the commons", *Science* 302 (5652): 1907–1912.
Folke, C., Carpenter, S.R., Elmqvist, T., Gunderson, L., Holling C.S., and Walker, B. (2002). "Resilience and sustainable development: Building adaptive capacity in a world of transformations", *Ambio* 31(5): 437–440.
Goodland, R. and Daly, H. (1996). "Environmental Sustainability: Universal and Non-Negotiable", *Ecological Applications* 6(4): 1002–1017.
Government of Botswana. (2007). Community Based Natural Resources Management Policy. W. a. T. Ministry of Environment. Gaborone, Botswana, Government Printer.
Hardin, G. (1968). "The Tragedy of the Commons", *Science* 162(3859): 1243–1248.
Holling, C. and Meffe, G. (1996). "Command and control and the pathology of natural resource management", *Conservation Biology* 10(2): 328–337.
Hunter Jr, M. L., Hitchcock, R.K., and Wyckoff-Baird, B. (1990). "Women and wildlife in Southern Africa", *Conservation Biology* 4(4): 448–451.
Leach, M., Mearns, R., and Scoones, I. (1999). "Environmental entitlements: dynamics and institutions in community-based natural resource management", *World Development* 27(2): 225–247.
Lebel, L., Anderies, J.M., Cambel, B., Folke, C., Hatfield-Dodds, S., Hughes, T.P., and Wilson, J. (2006). "Governance and the capacity to manage resilience in regional social-ecological systems", *Ecology and Society* 11(1): 19.
Levin, S. (2006). "Learning to live in a global commons: socioeconomic challenges for a sustainable environment",*Ecological Research* 21(3): 328.
Murphree, M. (2004). "Communal approaches to natural resource management in Africa: From whence and to where?" *Journal of International WildlifeLaw and Policy* 7(3–4): 203–216.
Mutandwa, E. and Gadzirayi, C.T. (2007). "Impact of community-based approaches to wildlife management: case study of the CAMPFIRE programme in Zimbabwe", *The International Journal of Sustainable Development and World Ecology* 14(4): 336–344.
Olsson, P., Folke, C. and Berkes, F. (2004). "Adaptive co-management for building resilience in social–ecological systems",*Environmental Management* 34(1): 75–90.
Ostrom, E., Burger, J., Field, C.B. and Norgaard, R.B. (1999). "Revisiting the commons: local lessons, global challenges", *Science* 284 (5412): 278–282.
Peterson, G. (2000). "Political ecology and ecological resilience: An integration of human and ecological dynamics", *Ecological Economics* 35(3): 323–336.
Phuthego, T. and Chanda, R. (2004). "Traditional ecological knowledge and community-based natural resource management: lessons from a Botswana wildlife management area", *Applied Geography* 24(1): 57–76.
Plummer, R. and Armitage, D. (2007). "A resilience-based framework for evaluating adaptive co-management: Linking ecology, economics and society in a complex world", *Ecological Economics* 61(1): 62–74.
Shackleton, C.M., Shackleton, S.E. and Cousins, B. (2001). "The role of land-based strategies in rural livelihoods: the contribution of arable production, animal husbandry

and natural resource harvesting in communal areas in South Africa", *Development Southern Africa* 18(5): 581–604.

van der Heiden, L.(1991). Land Use and Development Plan – Okavango and Kwando Wildlife Management Areas. Maun, Botswana, District Land Use Planning Unit and Tawana Land Board.

5 Partnerships and Power Games in Ecotourism Ventures of Indigenous Communities: The Case of Gudigwa Community in the Okavango Delta, Botswana

Lefatshe Innocent MAGOLE and Lapologang MAGOLE

Abstract

There is a now worldwide outcry that indigenous people around the world have suffered at the hands of more powerful members of the societies in which they live. In Botswana it is believed that *Basarwa* (San), the earliest inhabitants of the land, are in the periphery of the country's socio-economic development because of such marginalisation. As a community, the *Basarwa* are poor and politically disempowered. All sorts of organisations and individuals (NGOs, private entities, Government officials, politicians and international organisations) make decisions for them. Most devastating for these communities is that they are often barred from accessing natural resources such as land and wildlife, on which they depend more than any other community. This is despite the fact that often they live in close proximity to these resources. As a result, when ecotourism was proposed in the late 1990s as a pro-poor development approach, the *Basarwa* became targets of the 'experiment'. While the principles of ecotourism promised to fast track the long awaited socio-economic development of the poor and marginalised, the experience of some communities testifies differently. In this chapter we analyse the partnership dynamics of a *Basarwa* community in the Okavango Delta region of Botswana with an international non-governmental organisation (INGO) and a big tourism company in an ecotourism project. We conclude that the community ended being further marginalised and disempowered by their participation in the project when 'support' turned into 'control', more so because the Government of Botswana was completely absent from the scene.

Introduction

In Botswana tourism is seen as the emerging engine of economic growth through which the country will be able to diversify from mining and develop its under-developed rural settlements and communities. Ecotourism on the other hand is perceived as one of the few sustainable socio-economic development options for remote area dwellers (Stevens and Jansen 2002). These communities, although often marginalised and hence under-developed, reside in the most scenic parts of the country. In particular the minority ethnic groups of *Basarwa* (San) tend to live in relatively undisturbed and pristine areas, endowed with spectacular natural and cultural resources. It is these areas and the communities who reside therein that are often targets of the tourism drive with the belief that they will be empowered to emerge from their socio-economic and political exclusion, unemployment and hence poverty. To this end wildlife, wilderness and cultural heritage have been identified as Botswana's prime tourism assets (National Development 7 1991). Close to 40% of the country's land area is designated National Parks, Game Reserves and Wildlife Management Areas (WMAs). These are focal areas for the national tourism development drive.

In this chapter, we apply the stakeholder analysis approach to the study of the power dynamics of an ecotourism project which involved a *Basarwa* community as the 'owner' of the project, an INGO which acted as community 'mobiliser' and supporter (financial and technical), and a private safari company which acted as a 'marketing agent' for the tourism product on offer. Apart from the socio-economic profile of the community, we identify the actors, their interests or claimed stake in the project, their source and gravity of power and their means to achieve or safeguard their interests. We do this with the aim of assessing whether the partnership improved or worsened the position and situation of the Gudigwa (*Basarwa*) community as described below.

The concept of development

The impact of a country's development planning and implementation on different sections of society is often influenced by the ideology and concepts adopted by the planners. The evolution of development as a concept has ranged from emphasis on economic growth to sustainability and to development as a human right and choice issue (Sen 1999). Seers (1969), as quoted in Mabogunje (1989), conceived of development as involving not only economic growth but also conditions in which people of the country have adequate food and jobs and income inequality is greatly reduced. As he puts it:

> The questions to ask about a country's development are three: What has been happening to poverty? What has been happening to unemployment? What has been happening to inequality? If all three of these have declined from high levels, then beyond doubt this has been a period of development for the country concerned (Mabogonje 1989: 35).

Another disciple of Seers, Friedman recasts the questions and warns that "if one or two of these central problems have been growing worse, especially if all three have, it would be strange to call the result 'development,' even if per capita income doubled" (1992:1).

The development of society or community is seen as essentially a transformation of its mode of production. The mode of production refers to those elements, activities and social relationships which are necessary to produce and reproduce real (material) life. This Marxian conceptualisation of development as socio-economic transformation has three important implications. Firstly, development is emphasised as essentially a human issue – a concern with the capacity of individuals to realise their inherent potential and to cope effectively with the changing circumstances of their lives. Secondly, development involves the total and full mobilisation of a society. The task of changing the institutions in which the thinking (of the individual) finds expression cannot be undertaken in an ad hoc or piecemeal fashion. It has to be comprehensive and to invoke total political commitment; hence Myrdal's definition of development as "the movement of the whole social system upwards" (see Mabogunje 1989: 60). Thirdly, development represents a redefinition of a society's external relations. It involves a shift from an outward-orientated dependent status to a self-centred and self-reliant position with regard not only to the processes of decision-making (Friedman 1992), but more importantly the pattern and style of production and consumption (Mabugunje 1989).

It appears that Botswana has for a long time since independence in 1966 concentrated its development effort on economic growth and infrastructural development. This was fuelled by the mineral wealth which dominated the economy since the mid-1970s (Magole 2003). In the 1990s sustainability in terms of environmental conservation was embraced due to Botswana's participation in the Rio Declaration (Agenda 21), but also due to strong campaigns by international and local donor-funded environmental organisations such as the International Union for the Conservation of Nature (IUCN), Conservation International, Kalahari Conservation Society and the United Nations Environment Program (UNEP). During this period, and until recently, issues of social equity and access to resources have either been completely ignored, or only referred to and not implemented practically. This has had devastating implications for certain sectors of society, especially the *Basarwa*.

This chapter uses a *Basarwa* community to demonstrate the challenges of unequal partnerships in development which result from paying little or no attention to social issues as required by the concept of sustainability.

Development status of Basarwa

It is important to show the position of *Basarwa* in Botswana's socio-political and economic development arena, in order to appreciate the context within which the development of the *Basarwa* occurs. The *Basarwa* are disadvantaged due to both past and current discrimination. Politically, the *Basarwa* encounter social exclusion. To this end, their traditional leaders are often denied recognition, thus limiting their power and authority as compared with other *Batswana* groups (Hitchcock 1985). Often 'foreign' leaders or systems of leadership are imposed on them.

Furthermore, they have problems with the judicial system in terms of higher arrests, harsher sentences, and the inability to properly defend themselves because of language barriers. While there are no official cultural restrictions for *Basarwa* and other non-*Setswana* speaking groups, discrimination and marginalisation translates into such restrictions as these groups, especially the *Basarwa*, try to hide their identity in order to escape these social ills (Hitchcock 2002).

The *Basarwa* face de facto restrictions on their places of residence as they are more often likely to be relocated than other *Batswana* groups. Magole (2009) showed that most of the recorded conservation evictions happened to *Basarwa* communities. Those who remain on their traditional land still face the risk of resettlement and many are denied land rights because of the prospect of resettlement (Hitchcock and Biesele 2000). In general, the socio-economic status of people decline after resettlement, partly due to the cost of relocation, which is never adequately covered by compensation, and due to the amount of time it takes for new modes of production and sources of livelihoods to gain momentum. Furthermore, there are restrictions (without their consent) on what and how they can hunt, and they are often arrested and harassed as 'poachers' (Hitchcock and Biesele 2000). Basically, the government has limited what the *Basarwa* can get from the land to the point that they are not able to sustain a decent livelihood.

Unable to sustain decent livelihoods, alcoholism and associated social problems have become endemic in these communities. Sometimes resettled communities lack access to basic social services such as health care, education and clean water. *Basarwa* in general, and those of Gudigwa Village in particular, face socio-economic exclusion, unemployment, and poverty (Conservation International-Botswana 1999; Mbaiwa and Rantsudu 2003). As stated earlier, because the Gudigwa community is situated in a pristine wetland of international importance, the Okavango Delta, the community has been targeted for ecotourism development.

Given that ecotourism has been identified as a sustainable development option for residents of remote areas, the essence of any proposed development must adhere to the three principles of sustainable development: economic productivity and efficiency, social inclusion and equity, and environmental conservation. While environmental conservation and economic growth have historically enjoyed strong advocacy and acceptance, the social marginalisation of residents of remote areas has yet to receive the same level of attention.

Ecotourism

Ecotourism, as a concept, developed in the 1980s with the growth and expansion of global tourism, and increasing interest in travel to natural areas (Page and Dowling 2002). However, the concept has not been without controversy. The word 'ecotourism' has been so widely used that to some commentators it is almost meaningless in its current usage. According to Page and Dowling (2002), Ceballos-Lascurain has suggested that a lingering problem in any discussion on ecotourism is that the concept is not well understood and thus it is often confused with other types of tourism development.

An example of ecotourism lodges found in the Okavango Delta. Photo credit: Alyne Delaney

The nature of the debate on the definition of ecotourism is best summarised by Harrison (1997) below:

> In recent years ecotourism has become something of a buzzword in the tourism industry. To put the matter crudely, promoters of tourism have tended to label any nature -orientated tourism product an example of 'ecotourism' while academics have so busied themselves in trying to define it that they have produced dozens of definitions and little else (1997:75).

Guidelines also need to be produced for how ecotourism is to be implemented. Unfortunately while the debate on meaning has ravaged on, development of ecotourism principles has lagged behind, hence different players undertaking different tourism activities claim to be practicing ecotourism. Botswana has adopted The International Ecotourism Society (TIES) definition for ecotourism as outlined in the Botswana National Ecotourism Strategy (BNES) of 2002. Ecotourism is defined as responsible travel to natural areas that *conserves the environment* and *sustains the wellbeing of local people*. It is against this definition that we test ecotourism practices in Botswana. We take the case of Gudigwa Village and pose three questions: a) Did ecotourism help conserve their environment? b) Did ecotourism sustain their wellbeing (livelihoods and economic development)? c) Given the socio-political status of *Basarwa* in Botswana, did ecotourism empower the community to take control of conservation and management of their natural resources?

Ecotourism innovation and adoption in Botswana has been an evolving process. It started in the early 1990s as Community Based Natural Resource Management pioneered by the Department of Wildlife and National Parks (DWNP). As the process gained momentum and more stakeholders participated in CBNRM, Community Based Tourism was filtered out of CBNRM and referred to as ecotourism. The Department of Tourism then developed the National Ecotourism Strategy (NES) of 2002. The goal of the NES is to create an environment in which all elements of tourism development planning and management facilitate, promote and reward adherence to the key principles of ecotourism (Stevens and Jansen 2002).

Generally speaking, in Botswana the areas where *Basarwa* have settled are very remote and are usually furthest from the main administrative centres of the Districts. In terms of infrastructure provided by Government (roads, clinics, schools, etc.), *Basarwa* settlements are usually the least developed within the District and offer few employment opportunities. This is because they often live in small communities. The cut-off population for basic service provision in Botswana is 500 inhabitants. As a result, many *Basarwa* live below the Poverty Datum Line (PDL) and face difficulties in terms of access to social services, employment, and income-generating opportunities (Hitchcock and Biesele 2000).

On the other hand, the areas where *Basarwa* tend to live are relatively undisturbed and pristine areas, endowed with spectacular natural and cultural resources. These areas have become potential destinations for Ecotourism, which has been identified as one of the few sustainable development options available in these remote communities.

However, in order to participate and benefit from ecotourism a community needs the following:

- Moderate to high education levels among members of the community
- Creativity, innovation, and adaptability
- Financial resources
- Ecotourism products (e.g. scenic beauty and cultural diversity)
- Ecotourism support infrastructure
- Marketing capability.

There are at least three ‘exclusively’ *Basarwa* communities in the Okavango Delta involved in activities that can loosely be categorised as ecotourism – the communities of Mababe, Khwai and Gudigwa. Whilst the communities of Mababe and Khwai were engaged in safari hunting at the time of study, the Gudigwa community was engaged in photographic tourism and operated a photographic camp. All three communities have established Representative and Legal Entities (RALE) in the form of community trusts. The study presented in this chapter focused solely on the Gudigwa community.

The case of Gudigwa Village Ecotourism Project

The community formed a partnership with Conservation International (CI), an American- based conservation organisation, and Okavango Wilderness Safaris (OWS), a Botswana- based tourism company to establish an ecotourism venture. CI mobilised both the Gudigwa community and the donor community around the concept, and OWS assisted in formulating and marketing the product. The Gudigwa community was mobilised to form the Bukhakwe Conservation and Cultural Trust (BCCT), which represented and negotiated on behalf of the community. The community provided their land and aspects of their much sought-after San (*Basarwa*) culture and traditional skills for commercialisation. The final product was an ecotourism lodge/camp that offers accommodation, bush walks, and traditional San song and dance.

Data collection methods

The study was conducted between October 2004 and May 2005, and from March to June 2006. Data collection methods included document reviews, discursive interviews, household surveys and focus group discussions.

Literature review

Information on tourism in Botswana and ecotourism in general was collected from secondary sources (e.g., project reports, statistical bulletins, newsletters, articles, etc.). Baseline information on Gudigwa community was obtained from the Gudigwa PRA report of 1999, which was prepared by CI and the Mbaiwa and Rantsudu (2004) report on the socio-economic aspects of the BCCT of Gudigwa.

Discursive interviews

Discursive interviews were held with key informants who possessed a great deal of information about ecotourism activities in Gudigwa. This enabled researchers to further focus the study on relevant issues relating to development and tourism in this community. Semi-structured interviews were conducted using an interview guide. Key informants included community leaders (*Kgosi* or Chief, the Councillor, the VDC chairperson), leaders of the tourism ventures affecting the community (chairperson of the BCCT, chairperson of the Okavango Community Trust), representatives of all employment centres operating in the village (Government institutions and private enterprises), as well as staff of CI who provided technical and financial assistance for the Gudigwa camp.

Household surveys

Structured interviews were conducted, including a questionnaire administered to heads of households or their representatives, to collect data on socio-economic and cultural aspects of the community. In particular, information was collected

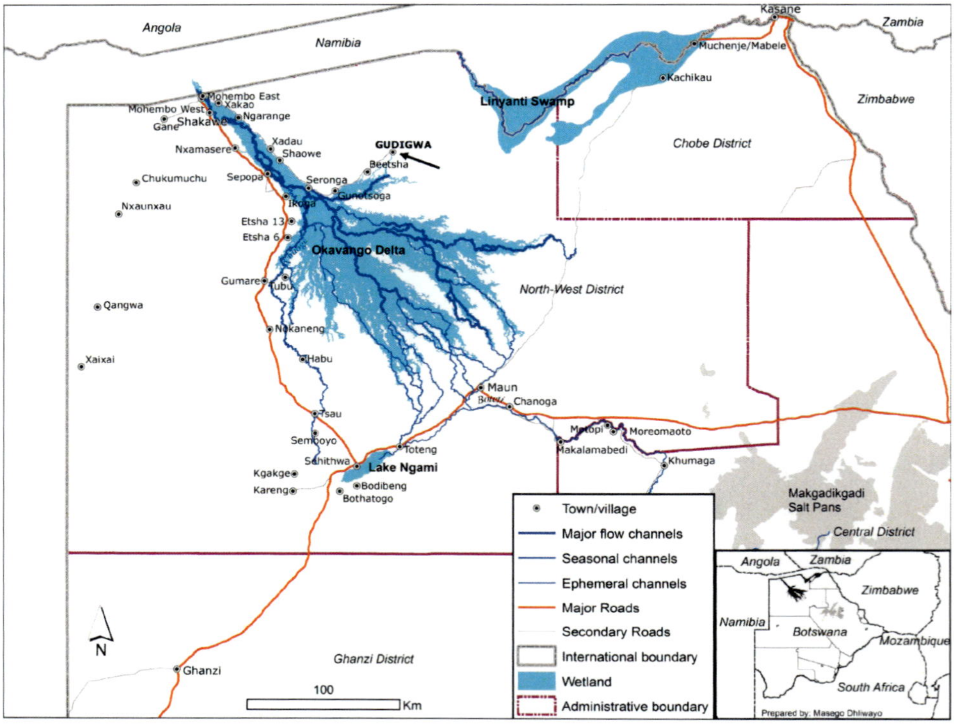

Figure 5.1: Map of Ngamiland showing settlement of Gudigwa. Source: ORI GIS Laboratory

on traditional skills possessed by members of the household and how they interact with the ecotourism establishment. All respondents were over 18 years old. Altogether 31 of about 72 household heads were interviewed. The choice of respondents involved stratified sampling of three wards in the village and random sampling of the households within the sampled wards.

Focus group discussions

Focus group discussions were held with the BCCT Board members, basket weavers and ostrich shell crafts groups and professional guides. Here again an interview guide was used to keep discussions on track and to ensure that all important issues were covered.

Data management and analysis

Qualitative data from key informant interviews and focus group discussions were analysed thematically to uncover emerging issues and identify themes. Data from the household surveys were analysed using Excel and are displayed in this chapter in the form of tables, graphs and pie charts to provide a visual depiction of the findings.

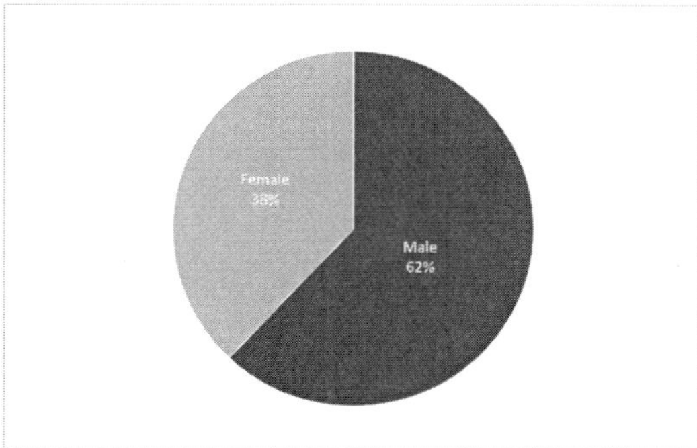

Figure 5.2: Gender stratification of household heads

The study area

GudigwaVillage is located on the north-eastern part of the Okavango DeltaRegion in Ngamiland District, Botswana (Figure 5.1). It is the last in a string of villages scattered along the Okavango River Panhandle that stretch from Mohembo East Village at the border with Namibia.

The village was established in 1987 through the country's Remote Area Development Program (RADP) as a service centre for nomadic groups. The inhabitants of Gudigwa are the San (*Basarwa*). According to the 2011 Housing and Population Census, Gudigwa has a population of about 725 people. The *Basarwa* of Gudigwa are composed of eight main clans that came from eight different areas – Xhondoro, Xharango, Gwakeqwe, Xhwatau, Ghicudza, Xhwakatsu, Hqwengu and Thobokhuru (Mbaiwa and Rantsudu 2004). Taylor (2000) notes that most of the people of Gudigwa came from //Gam/wi, and Letshaobe areas, not far from present-day Gudigwa.

Results and discussion

Livelihood activities of the people of Gudigwa include pastoral and limited dryland farming, gathering of veld products, employment in safari companies and the Government's drought relief projects. The community of Gudigwa has continuously attempted to diversify and expand its livelihood base. To this end the community has, with technical assistance from CI and financial aid from several donor agencies, formed a community trust, the Bukhakhwe Conservation and Cultural Trust and established an ecotourism camp, which opened officially in March 2003. The main aim of the camp was to enable the community to participate and derive socio-economic benefits from tourism in the Okavango Delta. GudigwaVillage is also part of the Okavango Community Trust (OCT), in conjunction with four other neighboring villages in pursuit of CBNRM projects.

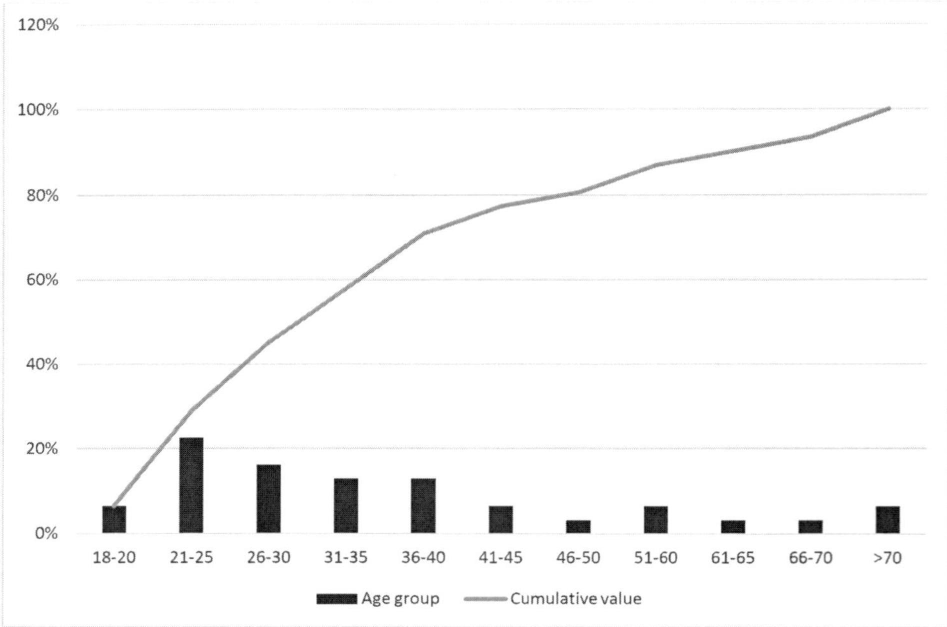

Figure 5.3: Age group structure (and cumulative) for household heads of Gudigwa

Profile of household heads

As shown in Figure 5.2, 39% of the surveyed households were female-headed and 61% were male-headed. In their 2003 survey, Mbaiwa and Rantsudu (2004) found the ratio to be 49% female-headed and 51% male-headed households. This differs from the national and district ratios where there are more (51%) (CSO, 2003) and increasing numbers of female-headed households. The assumption is that communities with a greater number of male-headed households are better off due to gendered differences in access to resources and livelihood opportunities which are biased towards men. However, this assumption does not hold for remote, isolated and underdeveloped areas where there are no employment opportunities even for men. Therefore, despite the fact that there are more – and increasing numbers of –male-headed households, poverty is still rife in Gudigwa.

39% of household heads in Gudigwa Village are aged 21–30 years (Figure 5.3). Furthermore, a large majority (71%) of household heads are aged 40 years and below. National statistics show that the 30–39 year age group is the dominant group (CSO 2004). Here also Gudigwa has an asset in young and able-bodied men remaining in the village. The experience in other rural areas is that this group migrates to urban centres leaving women, the young and very old to head households.

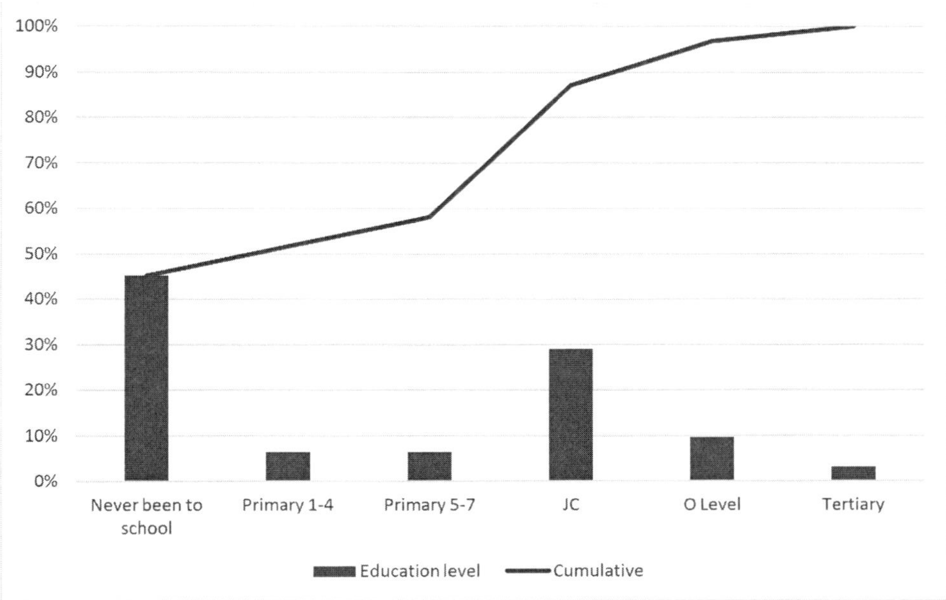

Figure 5.4: Education level of respondents (n=31)
Source: Magole and Magole 2007

Education

A high proportion (45%) of heads of households has never been to school whilst only 20% have attained O-level or above (Figure 5.4).This is comparable to the national rural level of 47% of heads of households having never attended school, but much higher than the national level where 30% of household heads have never been to school. Only 10% of those interviewed had any skills training. This may pose a problem for uptake when new and emerging livelihoods strategies like ecotourism are introduced.

Employment

Table 5.1 shows that at the time of the study there were 92 formal and informal sector employees at Gudigwa. 34% of jobs were provided by Government through the primary school, clinic and Tribal Administration, whilst the community based organisations (BCCT and OCT) provided 55% of jobs. The informal private sector (kiosks, local brewers) provided 11% of jobs. A sizable majority (74%) of the workforce in Gudigwa originates from Gudigwa. Further analysis of the jobs occupied by the locals shows that they are low paying jobs and therefore incomes remain low, resulting in minimal change to the poverty profile. Table 5.1 also shows clearly that the Gudigwa Ecotourism Camp provides most employment.

Figure 5.5 presents a profile of the traditional skills possessed by the people of Gudigwa. These are very important as indicators of the potential for self-

Table 5.1: Employment profile of Gudigwa Village

Employment centre	No. of Employees	No. of employees from Gudigwa
Primary School	18	8
Clinic	6	4
Tribal Admin	7	1
Vendor shop (4)	7	3
Local Brewer (3)	3	3
OCT	4	3
Gudigwa Camp	47	46
TOTAL	92	68

sufficiency by the community, but also identify skills that could be commercialised and utilised in a tourism venture. Indeed some of these were targeted by the project under study (see Figure 5.6). The most widely held skill is gathering, with all interviewees claiming to have this skill. The other popular skill is singing – claimed by 83% of the respondents. Other popular skills possessed by over 60% of the respondents are game tracking, game hunting, dancing and knowledge of medicinal plants. Scarce traditional skills which were claimed by less than 30% of the respondents include blacksmith, leather tanning, poetry and traditional healing.

It is evident that the community has the potential to become self-sufficient using their skills to make a living from their environment. This is how they have survived so long in a remote area. It is also unlikely that their livelihood systems were harmful to the environment as the condition of the Okavango Delta, their home for many decades, remains pristine. However since colonial times Government conservation policies have restricted their movement and access to resources, which has impoverished the community and created a level of dependency. Apart from many rules that make hunting almost impossible, the establishment of the Moremi Game Reserve closed off areas in which community members used to forage and gather wild plants for food and medicine. Livestock disease control fences have also restricted their access to natural resources. The settlement of Gudigwa is also surrounded by wildlife management areas (WMAs) (see Figure 5.1), which have many restrictions on the use of land resources. As a result, former artisanal and self-sufficient communities now live off aid from Government and other organisations. The community was also forced to consider the foreign concept of tourism as a source of livelihood, which meant that they had to depend on outside agents for the support required to establish and operate such a venture.

The profile of the traditional skills possessed by the people of Gudigwa is enormous. These skills are handed down from the old to the young through daily life experiences. Historically the skills were of great value for the daily activ-

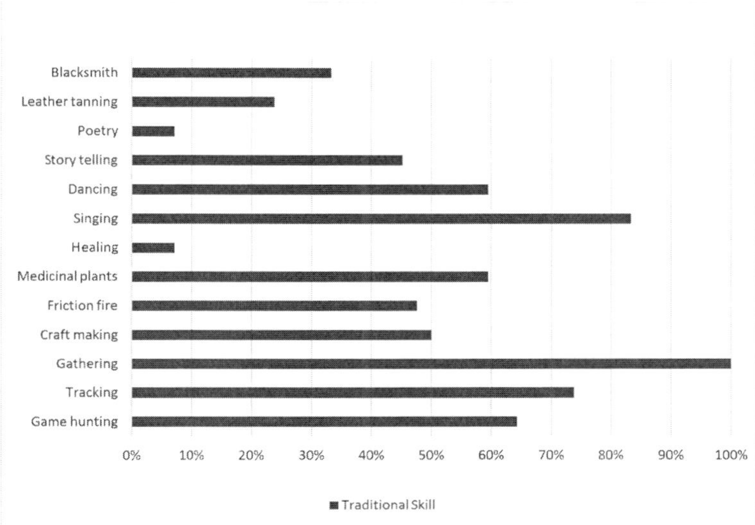

Figure 5.5: Knowledge of traditional skills by people of Gudigwa

ities of the people but in contemporary times these skills could be 'packaged' and marketed as cultural experience at the Gudigwa Camp (Magole and Magole 2007). However, Figure 5.6 shows clearly that many of the traditional skills that abound in Gudigwa are not utilised adequately or at all. Of 13 traditional skills identified, 85% (11) are underutilized – with gathering, tracking and game hunting highly under-utilized. According to key informants accounts and discussion in the focus groups, the design of the 'product' was handled by the marketing agent (OWS) and the technical assistance provider, CI, and did not consider the views of the community nor how they wished to use their skills and natural resources. The marketing agent in particular selected 'products' which are in line with their anti-hunting policies. According to members of the BCCT Board, the community's skill, culture and interest in hunting were rejected by the two powerful organisations.

Power dynamics in the project

According to members of the BCCT Board, from the very inception the process of establishing the project was characterised by unequal power relations, with the powerful OWS and CI dominating decision-making. OWS unreservedly pushed their business interests and secured exclusive marketing rights for the lodge, while CI pushed their preservationist conservation philosophy and refused the hunting safari option. Ultimately, the community conceded to the suggestions of CI and OWS, even though they would have preferred a different model, because these organisations raised the funds for the community activities. Here financial power was clearly at play and those who possessed it won.

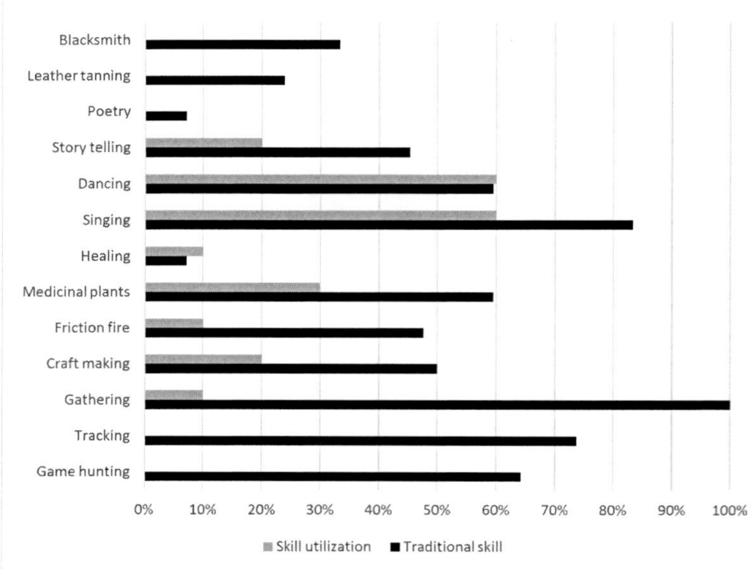

Figure 5.6: Utilization of traditional skills at the Gudigwa Camp
Source: Magole and Magole 2007

At the time when the study was carried out (2006), Gudigwa Camp was closed after having been burnt down for a second time. Anecdotal evidence suggests that the fire may have been arson. A few months prior to the fire, the camp was temporarily closed for a month because of a pending fight between the manager (the only non-Gudigwa employee at the camp) and one employee. While the marketing agent proposed the closure on the grounds of the safety of clients, the action made a clear statement to the rest of the camp employees and the community at large that such behaviour would not be tolerated and the community would be punished for it.

Mbaiwa and Rantsudu (2003) expressed the misgivings of the community about OWS's exclusive marketing rights for Gudigwa Camp. After reconstruction following the first fire in 2004, there was a protracted disagreement between the Gudigwa community and OWS over the marketing contract; OWS proposed a ten-year exclusive marketing contract and the community pushed for a shorter period. The BCCT Board, which was negotiating on behalf of the community, saw the proposed contract as disempowering to the community. The impasse was never resolved and although the community explored opportunities for other partnerships, the camp has remained closed indefinitely.

What is notable about this entire process is the absence of Government and its failure to take on its role as custodian of citizen rights and driver of community development. Was this because the matter involved a *Basarwa* (San) community who, as stated earlier, are at the bottom end of the social stratum of the country?

Or did Government stay away because the marketing company was viewed as an important investor whose reputation should not be tarnished by the 'Gudigwa incident'?

What is obvious is that due to past injustices, low education levels and low socio-economic status, the *Basarwa* are ill-equipped to participate as equal partners in ecotourism ventures with powerful stakeholders, and require political empowerment and support before any future participation in such ecotourism ventures.

Conclusion

The Gudigwa ecotourism case study lends itself to discussions of the power dynamics in development; the Gudigwa Camp operated both as a socio-economic development project and a disempowerment venture for the people of Gudigwa. Taylor (2000), quoting from Escober, presents development as a particular set of discursive power relations. He asserts that "thinking of development in terms of discourse makes it possible to maintain the focus of domination" (1996: 225). CI and other NGOs such as ACORD provided technical assistance as an 'intervention' in the desperate situation identified in Gudigwa through PRA methodologies. However, Twyman (1997) argues that such interventions are coercive rather than co-operative, reflecting a widespread prevalence of coercive conservation masked by 'community-based' approaches. The marketing contract between BCCT and OWS patronised the Gudigwa community, depicting the community as helpless and legitimising the OWS intervention. Taylor concluded that "the tragedy of CB-NRM in the northern sandveld is that it has so far failed to engage with the practices, priorities, aspirations and values of its potential beneficiaries" (2000: 276). He contends that the real beneficiaries are international tourists and conservation concerns rather than local residents.

While Gudigwa possesses, to a reasonable extent, the requirements for owning their means of production in an ecotourism enterprise, this is compromised by a lack of marketing capacity. Unless, and until, the community can undertake their own marketing, power relations will be biased towards the marketing promoter. Similarly, while CI has also played the role of technical advisor, the Gudigwa community must graduate from this relationship and develop their own creative, innovative, adaptable and educated human resources. It is here that we see the role of Government as important, through empowering its people and ensuring that they are protected from exploitative or fraudulent deals.

According to Mbaiwa and Rantsudu (2003), the people of Gudigwa have a deep hunter-gatherer culture that is in a continuum of acculturation; this study found the same. Mbaiwa and Rantsudu conclude that despite these problems, Community Based Tourism as characterised by Gudigwa Camp can offer a better livelihood strategy. In order to promote entrepeneurship and marketing skills,

Mbaiwa and Rantsudu recommend providing ongoing training for community members on tourism enterprise through workshops and short courses. Whilst this recommendation is plausible, it may not achieve the desired end result. The experience in older and more advanced community ventures such as the Okavango Community Trust is that once individuals acquire new skills, they often leave small remote villages for larger and more urbanised settings.

In terms of community empowerment and development, the Gudigwa Camp did not offer an optimum opportunity for the utilization of skills readily available within the Gudigwa community. As Friedman argues, "development is incapable of being objectively defined 'from the outside'" (1992: 41); in other words, the process must grow from within, with people building from a foundation of what is intimately known to them and expanding by acquiring knowledge from elsewhere. The process should induce the release of tremendous physical and mental energy with the certain practical outcome of socio-economic transformation required to launch a society or community on a path of self-centred, self-reliant and self-sustaining development. The pathway followed in the Gudigwa Camp ecotourism project portrays limited propensity towards these ideals.

It appears that in many instances, community projects have suffered from the hidden agendas of Government, donors, the private sector and NGOs, taking advantage of the unequal power relations between themselves and the local communities. Weak local community institutions have allowed elite capture of resource benefits by empowered actors at the exclusion of the masses. Furthermore, community projects generally have suffered from inadequate commitment of time and resources from most organisations purporting to promote community initiatives. From the Gudigwa case study, it emerges that there is a need for a dedicated, impartial, well-qualified and well-resourced facilitator with the sole role of mediating resource exploitation for local communities with other external parties. CBNRM requires a broker, along the same lines as the role played in the financial sector by a stockbroker, insurance broker, real estate broker etc. The role of the Natural Resource Management Broker (NRMB) will be to level the playing field between NRM players and broker win-win-win and sustainable deals. Taking seriously this role of a facilitator/broker may hold the key to the success and transformation of CBNRM.

References

Arca Consulting. (2000). Botswana Tourism Master Plan. Commission of the European Union and Government of Botswana.

Bolaane, M. (2004). "The Impact of Game Reserve Policy on the River BaSarwa / Bushmen of Botswana", *Social Policy and Administration*, Vol 38, No. 4, pp 399–417.

Central Statistics Office. (2004). *HouseholdIncome and Expenditure Survey 2002/2003,* Main Report Vol. 1. Government Printer: Gaborone.

Central Statistics Office. (2002). *2001 Population and Housing Census.* Population of Towns, Villages and Associated Localities. Government Printer: Gaborone.

Central Statistics Office. (2004). *Statistical Bulletin.* Quarter 4, 2001 and Quarter 1–4. 2002. Vol. 26 No. 4 and Vol. 27 No. 1–4 Government Printer: Gaborone.

Clark, M., Riley, M., Wilkie, E. and Wood, R.C. (1998). *Researching and Writing Dissertations in Hospitality and Tourism,* Thomson Learning: London.

Conservation International-Botswana.(1999). Gudigwa Participatory Rural Appraisal Report. Maun 20pp.

Friedmann, J. (1992). *Empowerment: The politics of alternative development,* Blackwell, Oxford.

Harrison, L. (1997). 'Ecotourism in the South Pacific: the case of Fiji', in *World Ecotour '97 Abstracts Volume*, Rio de Janeiro: BIOSFERA.

Hitchcock, R.K. (2002). "A Road to Self-Sufficiency: Natural Resource Management", *Cultural Survival*, Issue 26.1.

Hitchcock, R.K. and Biesele, M. (2000). "San, Khwe, Basarwa, or Bushmen?: Terminology, identity, and Empowerment in Southern Africa."www.kalaharipeoples.org

Hitchcock, R.K. (1985). "Foragers on the move: San survival strategies in Botswana parks and reserves", *Cultural Survival Quarterly* 9(1):31–36

Mabogunje, A.L. (1989).*The Development Process: A spatial perspective*, 2nd Edition, Unwin Hyman Ltd, London.

Magole, L.I. and Magole, L. (2007). "Is ecotourism the appropriate strategy for the socio-economic development of Basarwa (Khoisan)? The case of Gudigwa village, Okavango Delta, Botswana", In Allam Ahmed (ed.) *Managing Knowledge, Technology and Development in the Era of Information Revolution*, World Association of Sustainable Development, University of Sussex, Brighton. (pp 261–269).

Mbaiwa, J.E. and Rantsudu, M. (2003). "The Socio-Economic Baseline Survey of the Bukakhwe Cultural Conservation Trust of Gudigwa, Botswana." Harry Oppenhiemer Okavango Research Centre and Conservation International: Maun.

Page, S.J., and Dowling, R.K. (2002). *Themes in Tourism: Ecotourism,* Pearson Education Limited: Essex.

Statistics Botswana. (2012). "2011 Botswana Population and Housing Census, Alphabetical Index of Villages." Government of Botswana.

Stevens, P.W., and Jansen, R. (2002). "Botswana National Ecotourism Strategy." Government of Botswana.

Taylor, M. (2000)."Life, Land and Power: Contesting development in northern Botswana", PhD Thesis, University of Edinburgh, UK

Twyman, C. (1997). "Community Development and Wildlife Management: Opportunity and Diversity in Kalahari Wildlife Management Areas, Botswana", PhD thesis, Department of Geography, University of Sheffield, UK.

PART II – NATURAL RESOURCES MANAGEMENT AND USE ISSUES

6 Implementing a Management-Oriented Monitoring System as a Conflict Resolution Tool in Natural Resource Management in Tubu Village, Okavango Delta, Botswana

Lefatshe Innocent MAGOLE, Lapologang MAGOLE and
Mercy Happy MOEMEDI

Abstract

The BaYei are one of the ethnic groups that reside in the Okavango Delta, with temporary homesteads scattered throughout the floodplains. In the past they used to shift residence following the flooding of the Okavango River, depending on the time of the year and the livelihood activities undertaken at a particular time. As the livelihood activities supplied mainly subsistence needs, the impacts of extreme events such as drought or extreme flooding were minimal. Traditional natural resources management was led by the elders who controlled the harvest of resources and movement of livestock. A literature review and document perusal study showed that sedentarism as promoted by the Settlement Policy erection of veterinary fences and immigration by non-Yei settlers into the Okavango Delta are some of the factors that are diminishing the viability of traditional resource management. These factors also increase the severity of the impact of extreme events on the people as and when they occur. The study found that the answer may lie in a system of adaptive management where resource users have the capacity to monitor their resource use. To this end a Management Orientated Monitoring System (MOMS) was introduced in the Tubu community to monitor their resource use and improve their livelihoods. The results show that there is renewed confidence amongst community members to engage in their age-old livelihood practices and to formulate strategies that address challenges and seize opportunities. We conclude that by carefully merging and adapting new scientific knowledge and age-old livelihood practices, modern transformed and contemporary viable livelihoods can be established for the inhabitants of the Okavango Delta.

Key words: Livelihood, Production loss, Co-management, Management Orientated Monitoring System (MOMS).

Introduction

Natural Resource Management (NRM) deals with managing the way in which people interact with natural landscapes. Over the years there have been increased threats to the environment and livelihoods resulting in conflicts among users. In a bid to find solutions NRM approaches have evolved, as depicted in Figure 6.1.

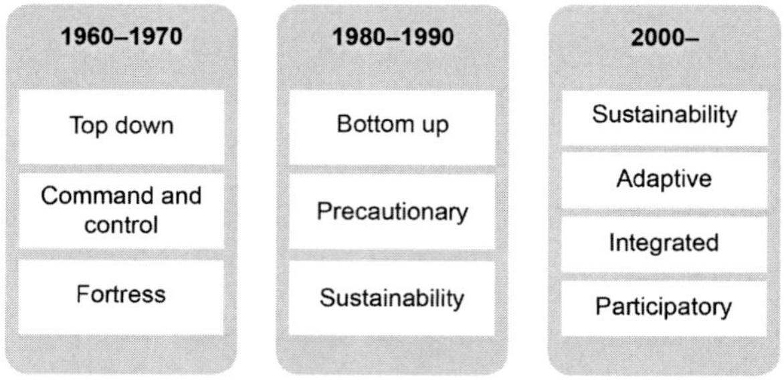

Figure 6.1: Evolution of NRM.

The first era (1960s–1970s) depicts Government control and policing of natural resources with resources users viewed as destructive and outdated in their harvesting methods and management systems. The second era (1980s–1990s) relied heavily on modern science and technical solutions to NRM issues. The current era (2000–) advocates for inclusive NRM systems which integrate all knowledge and stakeholders.

In the last evolution era, marked by integration and participation, monitoring of natural resources has become increasingly important as a way of measuring success or failure to meet the Millennium Development Goals (MDGs). Monitoring has been defined as the systematic measurement of variables and processes over time, and assumes that there is a specific reason for the collection of data such as ensuring that standards are being met (Spellerberg 2005). National legislative frameworks of many countries rely on an adequate knowledge of trends in species and habitats to make informed policy decisions, but monitoring receives low priority because it is expensive, and when it does take place it is usually externally driven (Danielsen *et al.* 2008). In particular there is a gap in local NRM monitoring, hence decisions are based on monitoring conducted by institutions and individuals who do not live with or near the resources.

In order to stimulate internal or locally based monitoring the Namibians formulated the ''Event Book System' (Stewart-Hill *et al.* 2005). The 'Event Book' was simplified deliberately and targeted for use at a local community level. The 'Event Book' System (also called Management Oriented Monitoring System or MOMS) differs from the conventional way of monitoring in that: (i) the community decides on what they want to monitor; (ii) the technicians only provide support upon request from the community and facilitate the design process; and (iii) all data collection and analysis is undertaken locally by the community members. To support local design, yet provide some national standardised and rigorous methodology, over time the monitoring system has been modularised by topic or theme. Topics/themes could be, for example, problem animals, poaching, vegetation, predators, fish, and so on. Once the community has selected what it wants to monitor, the technical support team provides a complete kit for each topic. Each kit contains 'tools' for: (i) data collection; (ii) monthly/quarterly reporting; and (iii) reporting and analysing long-term trends. These three different levels of data collection and analysis are usually colour coded to reduce confusion.

This chapter reports the results of piloting this monitoring system in the Tubu area, which is ridden with natural resource use conflicts between users, between sectors and between resources managers and users, that threaten the sustainability of both the resources and the community livelihoods.

Background

The Okavango Delta is a diverse, complex ecosystem with a wide range of resources and users, which are governed by multiple managers following an array of national laws, policies and guidelines, as well as regional and international conventions, agreements and protocols. Responsibilities for land use and resource management are dispersed across different sectors and are not coherently articulated. The development of the Okavango Delta Management Plan (ODMP) in 2008 (MEWT 2008) was the first attempt to integrate land and resource management within the Okavango Delta. Otherwise, land use management strategies exist within the various governing institutions – from the local to national levels – and these tend to be parochial and do not holistically address rural livelihoods, especially within a wetland setting.

The Okavango Delta is originally home to the main ethnic groups of Yei, Hambukushu, and Xanikhwe (River San). The Batawana and Baherero who live in the southern and western outskirts of the Delta are late arrivals, having arrived in the last century (Tlou 1985). These groups are dependent on the Okavango Delta for their livelihoods. Fishing, hunting, livestock grazing, floodplain cultivation and collection of plant materials for building, medicinal purposes, fuel and the production of handicrafts are important activities in the local economy (Kgathi *et al.* 2004, 2006). The main agricultural activities practiced in the Okavango Delta and its fringes are livestock farming and flood recession farming (Bendsen and Meyer 2003). These production systems are driven by the nutrient-rich soils found in the floodplains of the Okavango Delta.

The lifeblood of the floodplain is the annual flood pulse from the Angolan highlands which flows into the Okavango Delta, bringing with it nutrients and moisture in an otherwise nutrient-poor and arid system (Chimbari 2009; McCarthy 1998). These floods usually occur between June and August and thereafter begin to recede due to evapo-transpiration and infiltration (Ellery *et al.* 1998). The receding flood leaves behind suitable conditions for *molapo* farming (flood recession farming) (Murry-Hudson 2006; Chimbari 2009) and regeneration of nutrient rich forage for wildlife (Bonyongo 2004) and livestock. Thus, the availability of life-sustaining resources in the Okavango Delta shifts with the flood. This variability of the area's social-ecological system in both time and space presents challenges to management. Historically, the settlement and land use pattern in the area in and around the Okavango Delta was highly dynamic, being controlled by environmental factors such as the availability of potable water, changing flood levels, rainfall, tsetse fly distribution, and malaria (Murray-Hudson and Parry 1997: 31).

In recent years Government has developed policies and other regulatory measures to regulate the use and access to natural resources as per the 1970s–1980 NRM era. These have had two notable effects which are important for this study; they altered access to resources and excluded resource users from NRM. In 1982 the Veterinary Department erected a cordon fence running from the eastern sections of the Okavango Delta southwards and then again in the west, northwards up until NG25, (see Figure 6.2) to separate cattle from wildlife, particularly from buffalo which are known to be carriers of foot-and-mouth disease (FMD) without being affected by the disease themselves, and prevent the spread of wildlife-borne diseases such as FMD to livestock. Since the erection of the buffalo fence the likelihood of domestic stock coming into contact with wildlife has been minimised. However the erection of the fence, followed by the declaration of the Delta (planning zone 8) as a cattle-free zone (van der Heiden 1991), deprived local farmers of their traditional right to utilise the deep, inner Delta as a fallback grazing area. The local communities have generally respected the veterinary or 'buffalo' fence and even during the 1980s drought no attempts were made to forcefully bring livestock into the core areas of the Delta (Bensen and Meyer 2002).

However, conflict arose due to a gap in the fence and allocation of the area (NG25) just across from the community to a private concessionaire. Traditional areas of natural resource collection extend across the western buffalo fence into private concession areas NG24, 25 and 26. Villagers have been denied access to these resources on occasion and this has caused tension between locals and concessionaires. A problem analysis exercise found that the root cause of the conflict is the land use zoning decision made by the land authority and the subsequent land allocation without the knowledge of the community.

The fundamental problem with NRM revolves around the intersection between the science needed for environmental management decisions and property rights required for access. According to Ostrom (1991), property rights – which are complex bundles of access and ownership rights – are the single most important determinant of human behaviour in respect to the environment. This chapter presents the result of a pilot initiative aimed at involving the community resource users in the generation of the science needed to make NRM decisions in their area.

Study site description

The study was carried out in Tubu Village, 10km east of Gumare Village. The village is situated on the now-dry Thaoge River, on the western margin of the Okavango Delta. Livelihoods for villagers are still largely dependent on natural resources. The progressive drying of the Thaoge distributary has resulted in an increased intensity of use of the Karongana system to the east, partly within the WMA.

The demographics of the Tubu community are:

Population	483 (SB, 2012)
Age structure	70% older than 30 years of age
Number of households	226 (30% abandoned and or without occupants)
Employment	9 people have fixed employment
Subsistence dependant	96% depend on subsistence agriculture

(Source: Tubu Participatory and Integrated Land Use Management Plan 2010).

The settlement pattern can be divided into three areas. The first is the main village of Tubuwhere government has provided amenities in terms of a primary school, a clinic and a traditional administrative centre (*Kgotla*). The majority of the residents reside within this locality. Then there is an area for *molapo (*flood plain) fields, which extends along the western river system, where crop farming takes place. The main crop cultivated is maize. Finally there are the cattle posts, mainly spread along the western river system, where livestock (cattle, goats, donkeys and horses) is kept (see Figure 6.3). The entire area is approximately 200km^2 with a length of 20km and breadth of 10km.

Methodology

The study was conducted under the DARMA (2009–2012) project which followed the intervention of the BIOKAVANGO Project over a five-year period (2006–2010) in some of the 'hotspots' that were identified during the preparation of the ODMP. Tubu area was identified as the 'hotspot' of land use conflict in addition to high levels of poaching. The intervention process involved a stakeholder analysis, followed by a needs assessment and a facilitated negotiation process between and amongst stakeholders to develop a common understanding of the issues. This process was undertaken through Kgotla meetings, community workshops and key stakeholder interviews. In November 2007, all stakeholders with interest in the Tubu area agreed to establish a Joint Management Committee (JMC), made up of representatives from the local community, Government departments, the private sector and NGOs. The mandate of the JMC was to set up a forum at which issues concerning Tubu stakeholders could be discussed and negotiated. The JMC agreed

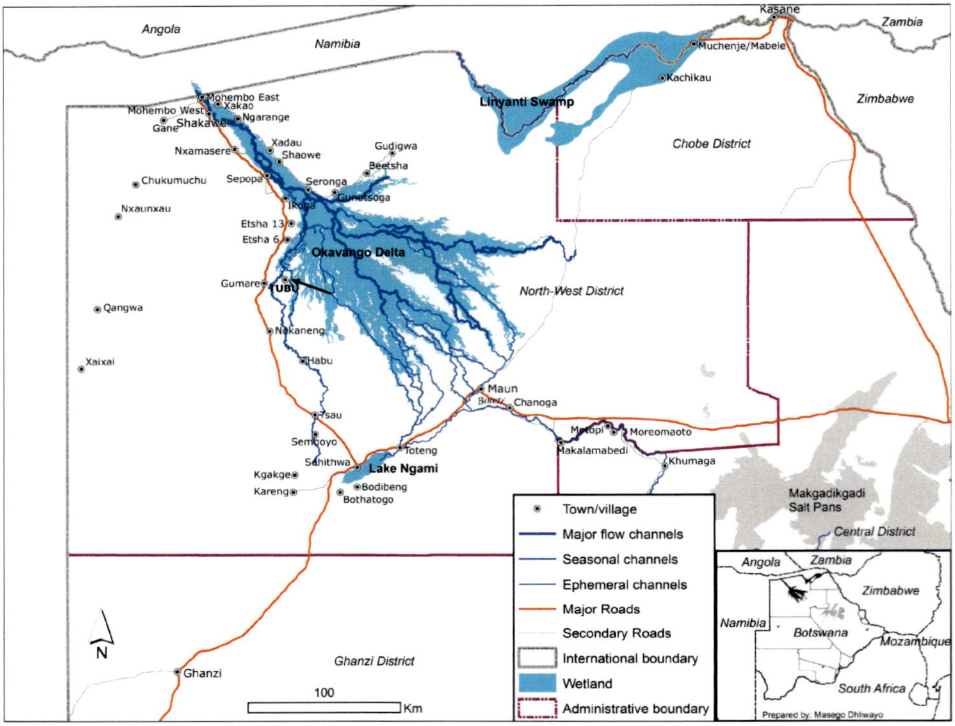

Figure 6.2: Map showing location of Tubu (study area). Source: ORI GIS Laboratory

to meet quarterly from January 2008 and to hold community meetings twice a year to give feedback.

The JMC agreed to adopt the Management Oriented Monitoring System (MOMS) to monitor resources and activities within their area. Five resource monitoring registers were developed covering: a) problem animals; b) *Molapo* farming activity; c) sighting of important wildlife species; d) livestock numbers; and e) veld resourceharvesting. Tubu conservation area was divided into six sub-areas (A–F) through a community mapping exercise (Figure 6.3). Twelve Community Rangers were selected by the community and trained over a period of three months on data collection or recording using the MOMS books. Ten Community Rangers were then engaged to monitor resources, activities and events in each sub-area from November 2009 to October 2012. Two of the trained Community Rangers were engaged as supervisors to supervise activities in three of the sub-areas.

A mapping exercise for Molapo fields and cattle posts was undertaken. The Molapo fields were mapped through GIS selection of ploughed fields in 2010 and 2011 as well as ground truthing. The process of mapping cattle posts involved visiting all posts, obtaining the GPS positions and recording the number of livestock at each post.

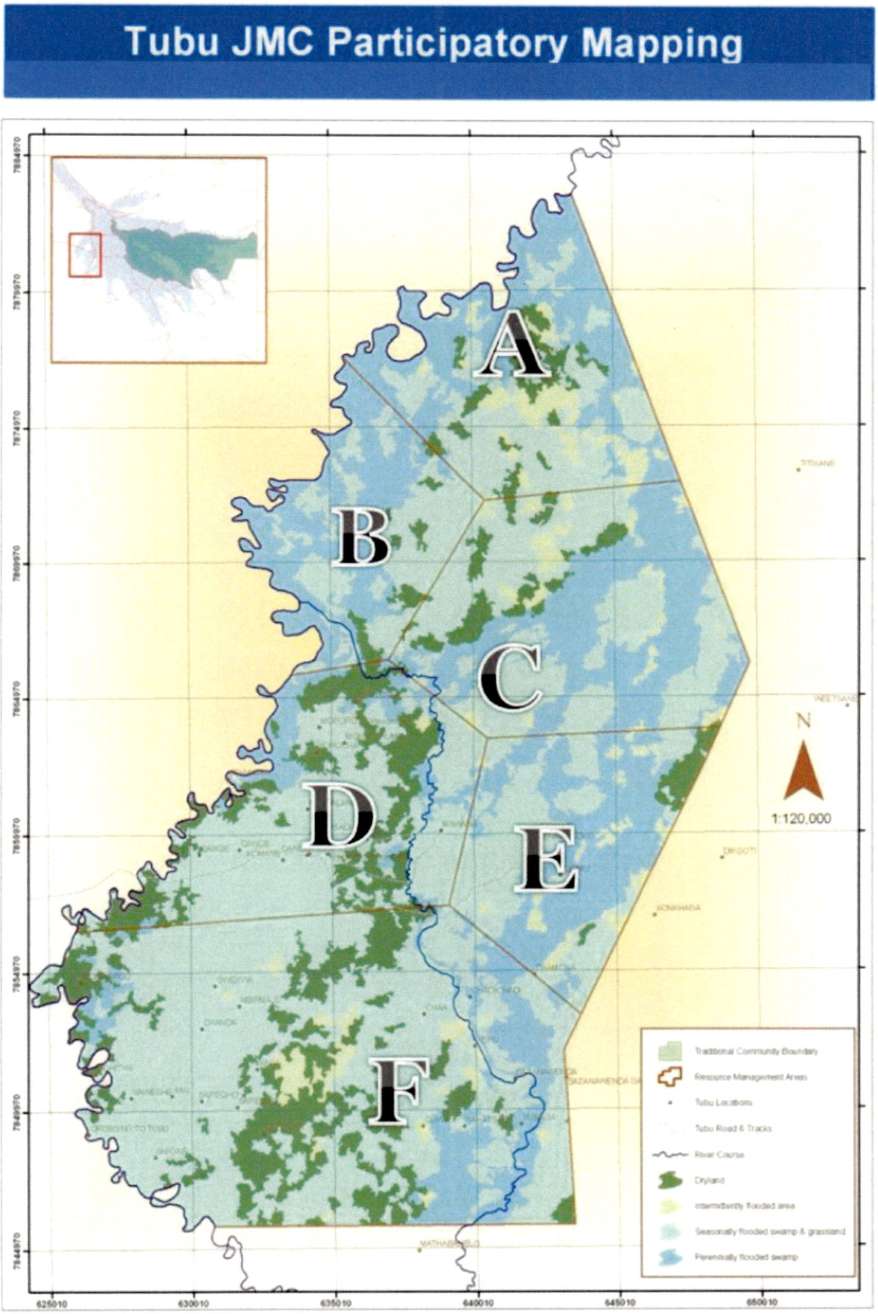

Figure 6.3: Delineation of monitoring areas. Source: BIOKAVANGO

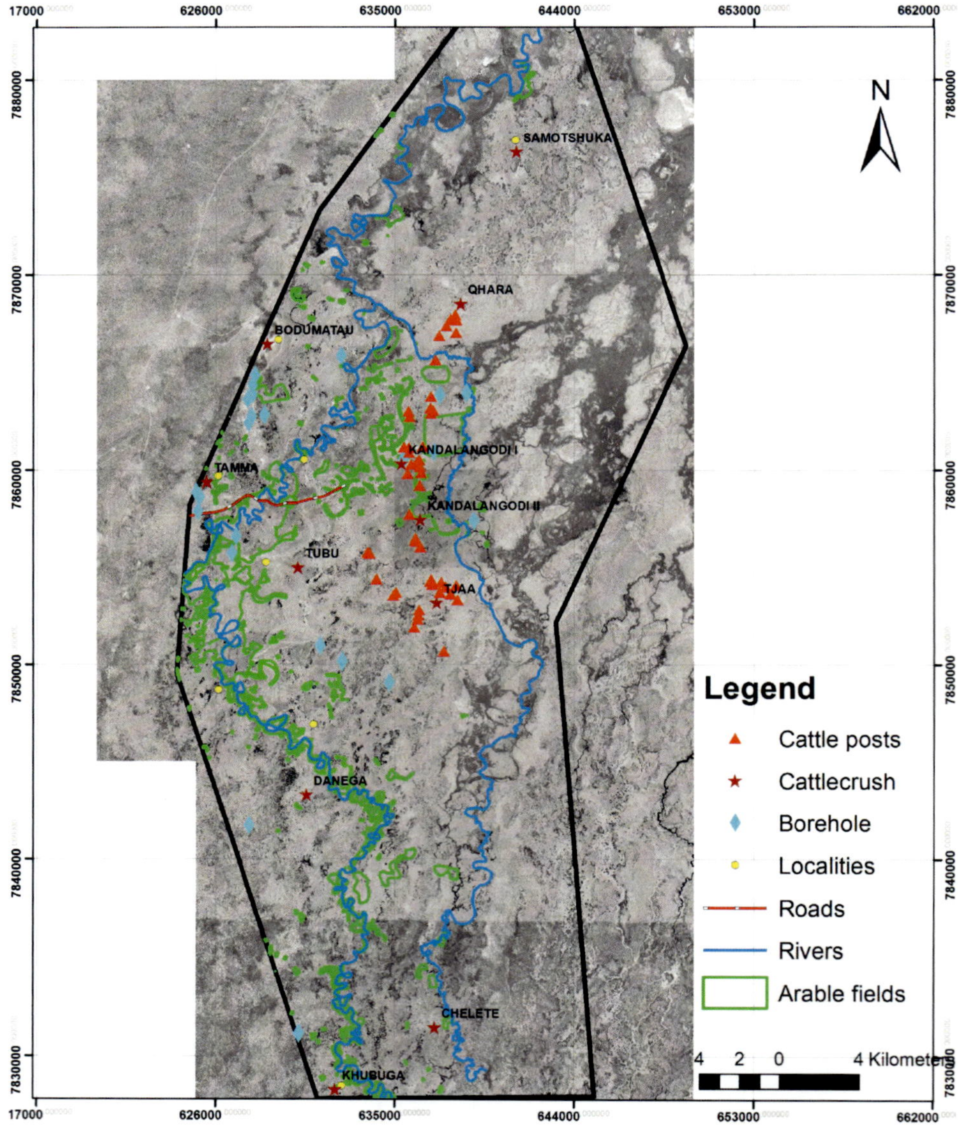

Figure 6.4: Distribution of cattle posts and molapo fields. Source: BIOKAVANGO

Meetings with local scientists and District officials were held once a year for the three years of the DARMA project tenure. Similarly, three annual community feedback sessions were held at the village *Kgotla* to disseminate the results and discuss new interventions based on the findings.

Results and discussion

Mapping of cattle posts and molapo fields

A total of 85 cattle kraals were recorded in the Tubu area. Each kraal consisted of cattle owned by members of the same family, and or close relatives. The number of cattle per kraal ranged from 3 to 450 head of cattle. Close to 65% of the kraals had less than 100 cattle and only 10% of the kraals had more than 200 cattle. All the farmers in Chaa, Mawana and Ngojo cattle posts indicated that they had two different locations where they move to depending on the state of their grazing resource and or extent of the flood. The farmers in Xomokao tended to be more sedentary.

From the results, it is estimated that the total number of cattle owned by the members of the Tubu community is between 5,000 and 6,000. However, the actual total herd number that is estimated to graze in the Tubu community area is in excess of 10,000 animals. It is understood that these additional animals are owned by either members of adjacent communities (Etsha 1–6 and Nokaneng) or by absentee owners who reside in urban areas in Botswana. These absentee owners are allowed by the Constitution to graze their livestock anywhere on tribal land in Botswana in accordance with the Tribal Grazing Land Policy (TGLP) and other legal instruments.

Trade in cattle was very small, with only six farmers reporting to have sold a single beast to other individuals over the three-year period. Whilst cattle were found exclusively at the cattle posts, goats were found at all the three land use zones (residence, fields and cattle posts). The flocks of goats ranged from 5 to 120 goats, with an average flock size of 20 goats. Goats were the most traded livestock, both for cash and bartering for other goods. Seventy six percent (76%) of the farmers reported having traded in goats in the three-year study period.

A total of 304 *molapo* fields were recorded from the GIS mapping. However, only 95 active *molapo* fields were found in the Tubu area during the mapping periods. Most (40%) of the fields were found in Area D (Mawana that includes the Tubu Village followed by area B – Ngojo (30%). The size of each *molapo* field ranged from 0.25 to 5 hectares; however, 97% of them were 2 hectares or smaller. Each family with a *molapo* field reported having at least one other *molapo* field downstream or upstream with some owning up to four other *molapo* farms. Only one *molapo* field is ploughed at any one time depending on the extent of the flood waters. The most popular crop grown is maize (planted by all farmers) followed by pumpkin, watermelon, sweet reed and beans respectively. All farmers broadcast their seeds and practice mixed farming. The most prevalent draught power was donkeys and cattle, whilst families with fields smaller than a hectare used hand-held hoes for ploughing.

Table 6.1: Identification of problem animals by year

	2010	2011	2012	Damage caused *(livestock killed or crop fields destroyed)*
1	Hyena	Hyena	Hyena	Goats
2	Lion	Lion	Lion	Cattle
3	Elephant	Elephant	Elephant	Crops
4	Porcupine	Porcupine	Porcupine	Crops
5	Crocodile	Crocodile	Crocodile	Donkeys, Horses and Cattle
6	Hippo	Hippo	Hippo	Crops
7	Waterbuck	Waterbuck	Waterbuck	Crops
8	Sitatunga	Sitatunga	Sitatunga	Crops
9	Martial Eagle	Martial Eagle	Martial Eagle	Kids (baby Goats)
10	Baboon	Baboon	Baboon	Crops
11	Buffalo	Buffalo	Buffalo	Crops
12		Reedbuck	Reedbuck	Crops
13			Jackal	Kids (baby Goats)
14			Wild Cat	Chickens
15			Civet	Chickens
16			Ant Eater	Crops
17			Python	Chickens
18			Slender Mon-goose	Chicken eggs
19			Steppe Buzzard	Chickens
20		Other (Cattle)	Other (Cattle)	Crops

Problem animals

Twenty different problem animals that cause damage to the community livelihoods were identified over the three years (see Table 6.1). Problem animals identified were wild animals and birds, while cattle was the one domestic animal identified as a problem. Damage caused was to crops in the fields at various stages of growth (from germination to harvest) and the killing of livestock (goats, cattle, donkeys, horses and chickens) by predatory birds and mammals.

More problem animals were identified in the second and third years of the study probably because the community became more informed about the study through feedback sessions and were willing to share more information with the Community Rangers. In addition, the Community Rangers could have improved their data collection skills over the years of the study.

Half (10) of the problem animals identified targeted crops. This suggests that crop production suffers multiple losses and several interventions are required to address this. Building fences or barriers around crop fields was the main intervention the farmers utilised to reduce losses and this was effective in protecting crops against most problem animals. However, elephants cause the greatest damage to

Cattle being driven from a maize field. This is a common occurrence due to weak or damaged fences around crop fields. Photo Credit: Alyne Delaney.

crops and cannot be stopped by fences. In addition, elephants destroy the fences and make access to crops easier for other problem animals. Chickens and goats were also targeted by four and three problem animals respectively. The crocodile was the only animal that was found to be versatile in causing damage, targeting multiple livestock (donkeys, horses and cattle).

In Figure 6.5, incidents of damage to crops and livestock are combined and there are two distinct periods in a calendar year when damage occurs. The most damage occurs from January to May and is a combination of both crop damage and killing of livestock. In the last three months of the year (October to December), there is another round of damage although it is largely restricted to livestock predation, with less crop damage occurring. Farmers in Tubu have two ploughing seasons: flood recession or *molapo* farming in August to September, and dryland or rain-fed farming during December to January.

The results of the study showed that the hyena was the most notorious problem animal, consistently recording over 30% of the damage incidents over the three-year study period, followed by the crocodile and the martial eagle. The lion and the elephant came fourth and fifth respectively, despite being singled out by most

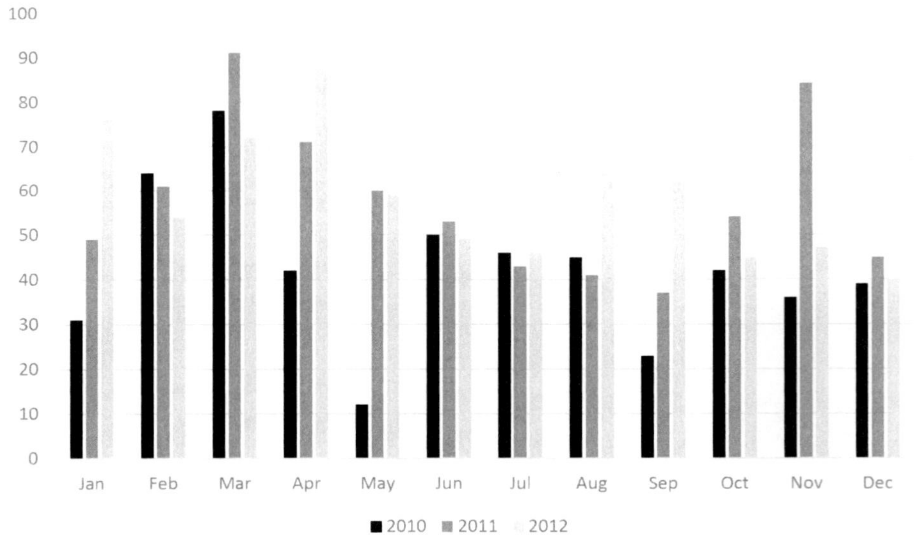

Figure 6.5: Number of reported problem animal incidents recorded per month over the period 2010–2012

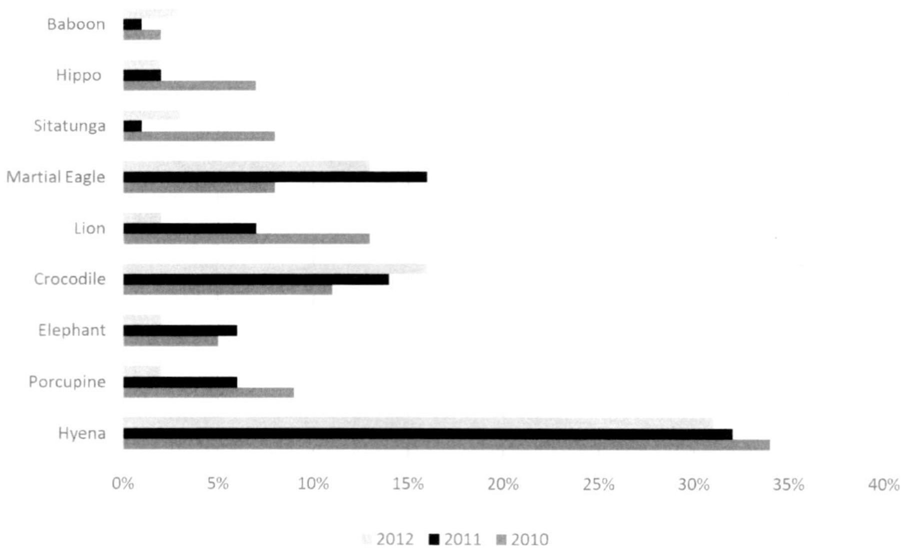

Figure 6.6: Proportion of damage caused by the top 10 problem animals for the period 2010–2012

stakeholders most often in discussions. This indicates the usefulness of collecting data (monitoring) in dispelling myths or solving misunderstandings.

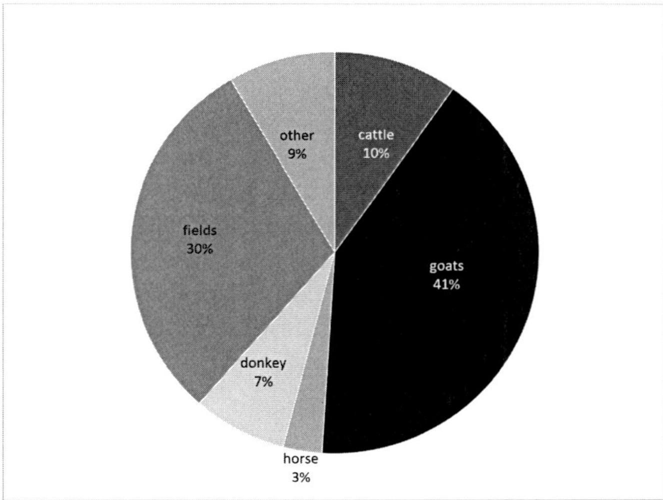

Figure 6.7: Proportion of species killed and or field damage incidents 2010–2012

Table 6.2: Number of livestock killed and estimated cost 2010–2012

	2010		2011		2012	
Damage	Number killed / damaged	*Estimate Value BWP*	Number killed / damaged	*Estimate Value BWP*	Number killed / damaged	*Estimate Value BWP*
Cattle	116	174,000	75	112,500	74	111,000
Goats	123	73,800	76	45,600	59	35,400
Goat Kids	119	35,700	35	21,000	56	33,600
Horses	42	84,000	24	48,000	18	36,000
Donkeys	63	31,500	17	8,500	7	3,500
Estimated Total livestock loss		*399,000*		*235,000*		*219,500*

Notes: Cost estimate of each domestic animal was based on market prices at Tubu in 2012, viz: cow = BWP1500, goat = BWP600, goat kid = BWP300, donkey = BWP500, horse = BWP2000. (US$1=P8). Cost estimates for crop damage were not computed since problem animals are only one of several factors that can result in crop failure (e.g. drought, pests etc.).

The livestock that experienced the most incidents from problem animals were goats, comprising 41% of the total damage incidents (see Figure 6.7). As shown in Table 6.2, goats are a target for hyenas, martial eagles and jackals and hence suffer the most losses. Damage to fields constituted 30% of all the damage incidents and was second after goats. Cattle loss incidents came in at 9%, despite local communities commonly mentioning cattle as the most affected by problem animals. Another myth was dispelled here. However in this regard it is worth noting that cattle are highly valued in Botswana in general and possession of a few cattle is considered great wealth. As a result, loss of one is considered a great loss when compared with loss of other livestock such as goats.

Table 6.2 provides a summary of the cost estimate of the cash value of damage caused by problem animals to the entire livestock industry of Tubu. The prices used were for the year 2012. Though in actual numbers goats suffered the most losses, the cost estimate for cattle is highest amounting to approximately BWP 400,000 (US$50,000) in the three years.

The annual livestock losses in cash cost estimates are high for a small rural subsistence economy such as that of Tubu. This, coupled with losses from crop damage, makes subsistence agriculture a risky livelihood.

Conclusion

The Tubu community is largely dependent upon subsistence forms of livelihoodswhich are intricately connected to periodic floods that occur in the area. Previously, the community of Tubu used to move with the flood in circular movements within the area. They have had to become more sedentary as result of new settlement policies. Consequently if they are not able to continue with their localised migration, the community becomes vulnerable to disasters such as drought, flooding and pandemic diseases – both human and animal related – as well as exposure to problem animals. Flooding influences a wide range of human activity in the area, ranging from when *molapo* crops are planted and reaped, when certain natural resources (reeds, grass, roots, tubers etc.) are harvested, where cattle are grazed and the movement of predators in search of food.

The losses sustained by the Tubu community due to problem animals make subsistence agriculture a high risk livelihood. The community seems to have accepted the losses as part of their livelihood system. To reduce the losses, deliberate Government policies need to be enacted that are informed by the situation on the ground. For instance, whilst the hyena is the most notorious problem animal, the Government compensation scheme does not cover it. With this data in hand the community can negotiate for the hyena to be covered in the compensation scheme or work together with other stakeholders on a solution to alleviate the losses and resultant poverty caused.

The local communities practicing livestock and *molapo* farming in the Okavango Delta have over the years developed their own strategies to cope with natural variability. Rashen (1988) points out that farmers usually move in response to flooding and in some cases one farmer will have access to several fields located in different areas, and hence plant in fields with suitable moisture conditions. Ownership of productive land in several places provides an opportunity for local communities to take advantage of different micro-climatic conditions in order to spread the risk (Ellis 2000). This strategy is difficult to implement under the current policy and supporting legal statutes governing land allocation and tenure in Botswana as they promote sedentarism. We conclude that these policies are flawed on two counts; they are not informed by a robust and holistic local situation analysis and they lack input from local stakeholders.

The pilot of MOMS as a resource monitoring tool has generated reliable scientific data, collected by the local communities, and confirms that rural communities have the capacity to generate scientific knowledge. MOMS offers rural communities a bridge to engage in the NRM discourse with other actors (such as Government, NGOs, academia). Under the unequal power relations between Government and rural communities, MOMS offers rural communities 'some clout' to contribute information towards NRM decision-making processes.

The MOMS exercise as whole was also empowering for the community. After many years of being excluded from NRM, even in their own local area, the community had grown to accept that they have nothing to contribute in the form of knowledge or ideas towards solving problems. Through the MOMS exercise, they learned a lot as some of the myths and beliefs they had held for a long time were dispelled by their own data. This is the essence of sustainability and adaptive management which encourage collective action in all stages of NRM.

References

Berkes, F. (1989). Common Property Resources: Ecology and Community Based Sustainable Development, London and New York: Belhaven Press.

Bendsen, H. and Meyer, T. (2003). "The dynamic of the land use systems in Ngamiland, Botswana: Changing livelihood options and strategies", Paper presented at Environmental Monitoring of Tropical Wetlands, Maun, Botswana.

BIOKAVANGO Project. (2009). Tubu Community Conservation Area Management Plan 2009–2018, University of Botswana, Maun.

Bonyongo, C. (2004).The ecology of the large herbivores in the Okavango Delta Botswana, PhD thesis, School of Biological Sciences, University of Bristol, UK, 125 pp.

Chambers, R. and Conway, G. (1992). Sustainable Rural Livelihoods: Practical Concepts for the 21st Century, IDS Discussion Paper 296, IDS, Brighton.

Darkoh, M.B.K. (1996): "Towards an Adaptive and Community-based Approach to Management of Natural Resources in Drylands in Sub-Saharan Africa", In Hjort-af- Ornas A. (ed) *Approaching Nature from Local Communities: Security Perceived and Achieved,* Motala: Motala Grafiska pp. 73–99.

Ellis, F. (2000). *Rural livelihoodsand diversityin developing countries,* Oxford: Oxford University Press.

Hitchcock, RK. (1985). "Water, Land and Livestock: The Evolution of Tenure and Administration Patterns in the Grazing Areas of Botswana", In L.A. Picard (ed.)*Politics and Rural Development in Southern Africa: The Evolution of Modern Botswana,* London: Rex Collins Ltd.

Holling, C.S and Meffe, G.K. (1996). "Command and Control and the pathology of natural resources management", *Conservation Biology,* Vol. 10, No. 2, pp 328–337.

Kgathi, D.L., Kniveton, D., Ringrose, S., Turton, T., Vanderpost, C., Lundqvist, J., Savanije, H., Seely, M., el Obeid, S., and Mendelson, J. (2006). "The Okavango; a river supporting its people, environment and economic development", *Journal of Hydrology*, Vol. 331, Issues 1–2, November 2006.

Kgathi, D. L., Bendsen, H., Blaike, P., Mbaiwa, J., Ngwenya, B. and Wilk, J. (2004). "*Rural Livelihoods, Indigenous Knowledge Systems, and Political Economy of accessto natural Resources in the Okavango Delta, Botswana*", Maun: University of Botswana.

Magole, L. (2003). "Tragedy of the Commoners: The evolution of communal rangeland management in Kgalagadi, Botswana", PhD Thesis, School of Development Studies, University of East Anglia, UK.

Magole, L. and Magole, L.I. (2009). "The Okavango; Whose Delta is it?" *Physics and Chemistry of the Earth*, Vol. 34, Issues 13–16, pp 874–880.

Magole, L.I. (2009) "Common pool resources among San communities in Ngamiland, Botswana", *Development Southern Africa*, Vol. 26, Issue 4, pp 597–610.

McCarthy, T. S. and Ellery, W.N. (1998). "The Okavango Delta", *Transactions of the Royal Society of South Africa* 53: 157–182.

Ministry of Finance and Development Planning. (2002). Revised National Policy for Rural Development, Government Paper No. 3, Republic of Botswana. Gaborone: Government Printing.

Ministry of Local Government. (1986). National Settlement Policy. Republic of Botswana. Gaborone: Government Printing.

Ministry of Local Government. (2004). Revised National Settlement Policy. Republic of Botswana. Gaborone: Government Printing.

Ministry of Local Government. (2004). Ngamiland District Settlement Strategy (NDSS), Final Report, Ministry of Lands and Housing, Department of Town and Regional Planning.

Rashen, K. (1988). "Economic findings and results of dryland and molapo farming systems of Western Ngamiland. Agricultural Development Ngamiland," Technical Report No. 5, Department of Agricultural Research, Ministry of Agriculture, Maun

Scoones, I. (1998). "Sustainable Rural Livelihoods: A Framework for Analysis", IDS Working Paper 72, IDS, Brighton.

Steiner, A. and Rihoy, E. (1995). "The Commons Without Tragedy? Strategies for Community-based Natural Resources Management in Southern Africa", Proceedings of the Regional Natural Resources Management Programme Annual Conference: SADC Wildlife Technical Co-odinating Unit, Kasane, pp. 1–36

Taylor, M. (2000)."Life, land and power, contesting development in northern Botswana", PhD thesis, University of Edinburgh.

Tlou, T. (1985). *History of Ngamiland: 1750–1906 The Formation of an African State,* Gaborone: Macmillan Publishing Company.

Twyman, C. (2000). "Participatory Conservation? Community-Based Natural Resource Management in Botswana", *The Geographical Journal*, 166 (4): 323–335.

Twyman, C. (1998). "Rethinking community resource management: managing resources or managing people in western Botswana", *Third World Quarterly*, Vol. 19, No. 4, pp 745–770

7 Tourism Planning and Development in the Okavango Delta: Are Communities a Misfit?

Joseph Elizeri MBAIWA

Abstract

This chapter demonstrates the phenomenal expansion of the tourism sector in the Okavango Delta and questions its sustainability. The primary research question is: What is the impact of this growth on the ecosystem, the economy and the people of Ngamiland District? Using data collected from Government documents and past research, as well as interviews with residents of some communities in the Delta, great success has been achieved in policy implementation and tourism development in the Okavango Delta. Tremendous progress has been made in planning for tourism development both at national and local levels. There has been substantial infrastructural development to support the industry, and over the years it has attracted an increasing number of clients. There is evidence of growing financial benefits to the national economy. Potential benefits to communities have also been revealed through the case studies of, Khwai and Mababe Villages that have in the past few years enjoyed benefits from implementing CBNRM projects. It is not clear, however, if the system is prepared and ready to address the potential negative socio-cultural, economic and environmental impacts of such high growth in an area with the level of ecological sensitivity of the Okavango Delta. The chapter concludes with the cautionary note that the speed of development may cause the institutional system to compromise on important principles of sustainability such as co-management and exclude local people from policy development, hence limiting their access and benefits from the natural resource of the Okavango Delta.

Introduction

Wildlife and – to a lesser extent – wilderness, are Botswana's biggest tourist products. Wildlife-based tourism in Botswana is carried out in national parks and game reserves. Most of these protected areas are located in the northern parts of the country in areas such as the Okavango Delta and Chobe region. This study was carried out at a time when international tourism in wilderness areas considered 'unspoilt and pristine' in developing countries is on the increase. The Okavango Delta is one such international nature-based destination in Africa, largely known for its rich natural resources such as forests, wildlife, water, birds, reptiles and others. The availability of permanent water resources in the Delta is critical for the survival of these species. In the past five years, more than a quarter of a million tourists experienced the incomparable beauty of the Okavango Delta, almost double the number of the previous five years. Growth in tourist numbers, facilities, infrastructure, aircraft operations, and tourism services have led to a booming tourist economy built around what is perceived internationally as a 'new' and 'exotic' destination. The question addressed in this chapter is how has the socio-economic situation of the Okavango Delta riparian communities changed? The answer lies in the management concept and resulting system that has been adopted for the exploitation of the Okavango Delta natural resources.

There is general consensus among scholars and practitioners that sustainable development is the ultimate goal, and different aspects of it help achieve specific objectives. Arguably the specific objective of rural resource conservation, sustainable livelihoods and poverty alleviation in natural resource dependent communities requires co-management approaches to NRM in line with the principles of sustainable development.

Co-management

The co-management/joint management principle can be equated to the sustainable development concept and the ecosystems approach. The basic principle in these approaches is that the management of natural resources should be people-centred in order to achieve resource sustainability in biodiversity rich areas. A common understanding put forward by Borrini-Feyerabend (2000) defines co-management as a situation in which two or more social actors negotiate, define and guarantee amongst themselves a fair sharing of the management functions, entitlements and responsibilities for a given territory, area or set of natural resources. In this regard, co-management of natural resources is assumed to be one of the ways to achieve sustainability in resource use in biodiversity rich areas. Co-management of natural resources across the world is designed to achieve the following objectives: i) identification of a common and agreeable definition of co-management in natural resources with particular reference to wildlife, birds and their interactions with culture and society and so on; ii) identification of the different forms, concepts and approaches of co-management in natural resource management.

In some contexts co-management has also been referred to as the joint management system. The term 'joint management' in this context means the establishment of a legal partnership and management structure which reflects the rights, interests and obligations of the local communities living within or around a protected area (or national park in the Okavango case), as well as those of the relevant Government departments, acting on behalf of the wider community. Joint management arrangements represent a trade-off between the rights and interests of traditional owners and the rights and interests of government conservation agencies and the wider community.

Joint management or co-management approaches represent a shift away from the protectionist/preservationist approach used in the management of national parks and game reserves especially in developing countries in that they allow a bottom-up management approach as compared to a top-down management approach for protected areas. In a bottom-up approach, different level stakeholders participate in resource management as appropriate or as decided by the parties, whereas in a top-down management approach, only the central or national government institutions control or manage the protected resource without input from local communities or any other stakeholder with interest in the resource. In addition, joint management allows local communities to derive socio-economic benefits from the resource as compared to the protectionist approach which does not allow communities living within or around protected areas to derive benefits from these areas.

Communities are social actors in themselves and provide the most natural and effective unit of identity, integration and defence for many under-privileged groups and individuals (Farvar 1999). Yet, communities are not homogenous entities, and their internal subdivisions should be recognised. In other words, while keeping their basic cohesion and identity, a plurality of values, interests and concerns should be recognised *within* any local community that participates in local resources management (Agrawal 1997).

As indicated earlier over the past two decades the Okavango Delta has become an important international tourist destination. However, not much assessment has been done to determine the extent to which local communities have been involved in the development and implementation of the NRM policies, especially those relating to tourism development. It has not been established either whether principles of co-management are applied in tourism development in the Okavango Delta. Furthermore there has been no comprehensive study to understand how the Okavango Delta communities benefit from current tourism development and investment. To shed light on these, this chapter will:

- Examine how natural resource policies and plans will affect the nature of tourism development in the Okavango Delta
- Review literature on the different type of tourism activities dependent on water resources
- Present a study of the tourism product and other dynamics in the Okavango Delta such as trends, origins and types of touristsvisiting the area
- Examine the impact of tourism development on the livelihoods of Okavango Delta communities
- Review the level of development of tourism infrastructure within the Okavango Delta and how it has influenced development in the District in general.

National policies and plans

There are several national policies and development plans that affect tourism development in Botswana in general and in the Okavango Delta in particular. These include:

- The Tourism Policy of 1990 (under review)
- The Wildlife Conservation Policy of 1986 (under review)
- The Community Based Natural Resource Management Policy of 2007
- National Ecotourism Strategy of 2002
- Ngamiland Tourism Development Plan of 2007

The Tourism Policy of 1990

The key policy for promotion of tourism in Botswana and the Okavango Delta is the Tourism Policy of 1990. It describes tourism as the new 'engine of growth' with the main aim of diversifying the country's economy from reliance on diamond mining through the promotion of tourism (GoB 1990). Some of the specific objectives of the policy include:

- To increase foreign exchange earnings and government revenue
- To generate employment, mainly in rural areas
- To raise incomes in rural areas in order to reduce urban drift
- To promote rural development and stimulate the provision of services in remote areas of the country.

The Tourism Policy also emphasises the need for increased local participation in the tourism industry. It acknowledges the need for diversification of the tourism product in order to reduce pressure on the natural resource attractions in the northern part of the country such as the Okavango Delta and the Chobe National Park. This Policy is currently under review to bring it in line with current thinking in tourism development. It is hoped that the review will promote diversification in the development of tourism. Expected long-term impacts of the reviewed Tourism Policy in the Okavango Delta also include an increase in tourism activities, infrastructural development and a rise in tourist numbers. Less desirable will be the increased pressure on natural resources and the attendant environmental impacts, which will need to be managed to ensure that the wetland resources are not degraded.

The other scenario for the Policy to consider and address is that if flooding levels in particular parts of the Okavango Delta change (e.g. decrease), either through the natural process or through human interference upstream (e.g. Namibia or Angola), there is likelihood that tourism development in the wetland will be affected.

The Wildlife Conservation Policy of 1986

The Wildlife Conservation Policy of 1986 is frequently seen as the blue-print for the re-introduction of community involvement in wildlife conservation through the implementation of Community Based Natural Resource Management Projects (CBNRMP). The Wildlife Conservation Policy is the precursor of virtually all the changes in wildlife conservation currently being carried out. It was designed to promote economic development of rural areas through the implementation of community tourism projects. The policy recognises the potential value of both consumptive and non-consumptive use of Botswana's wildlife resources by people living in wildlife areas. The overall aims and objectives of the Wildlife Conservation Policy are:

- To develop a commercial wildlife industry in order to create economic opportunities, jobs and incomes for the rural population and to enable more rural communities to enter the modern wage economy. This is hoped to reduce the number of rural people relying on subsistence hunting.
- To implement rational and effective conservation and management programs that will ensure that wildlife is utilised on a sustainable basis.
- To obtain good economic return on the land allocated for wildlife utilisation.

Through the Wildlife Conservation Policy of 1986, land in wildlife areas was sub-divided into zones known as Controlled Hunting Areas (CHAs). CHAs are administrative blocks or land units used by the Department of Wildlife and National Parks (DWNP) to allocate wildlife quotas to local communities for tourism purposes. After allocation of a CHA, a community decides on the type of land use for that particular area, that is, either for hunting or photographic purposes. In communities that use their CHAs for safari hunting, most of them sub-lease to safari hunting companies with such knowledge and skills. Most of these are international companies from developed countries.

It can be argued that the government policy to zone wildlife areas such as the ones around the Okavango Delta into CHAs for community-based tourism purposes indicates a partial return of land and its resources to local communities after their dispossession a century ago under British colonial rule. Access to land for community-based safari hunting purposes can therefore empower local people and improve rural livelihoods and sustainable wildlife resource use in the Okavango Delta.

The CBNRM Policy of 2007

The CBNRM Policy of 2007 recognizes that "rural livelihoods in Botswana are today less dependent on arable and livestock farming-based agriculture than in the past. 47% of all households cannot meet their basic needs" (Ministry of Environment, Wildlife and Tourism, 2007: 1). As a result of the problem caused by stunted agricultural development, Government aims to diversify the rural economy through natural resource development in order to alleviate rural poverty and sustain livelihoods. This policy outlines the governmental view, position and action with regard to the relation between natural resources, livelihoods and rural development. It declares that:

> People who live closest to natural resources generally absorb the greatest costs associated with their conservation. Given proper awareness and incentives, they are most likely to successfully benefit from and conserve such natural resources within their environs. For communities to actively engage in natural resource conservation, the benefits from such resources must exceed the costs of conservation. CBNRM aims to achieve this by offering eligible communities opportunities to earn tangible benefits from sustainable natural resources management (Ministry of Environment, Wildlife and Tourism, 2007: 1).

In addition to the drive to diversify the economy and sustain rural livelihoods through CBNRM, the Government also aimed to tackle issues currently facing CBNRM. The CBNRM Policy of 2007 further states that:

> CBNRM across the country is taking place without the guidance of a formal policy. As a result the dynamics of power relationships and personal interests in some communities has jeopardized the welfare of the wider community. Inappropriate financial management practices have prejudiced the wellbeing of some communities. Although numerous private enterprises and Community Based Organizations (CBO) have established mutually beneficial partnerships, in most cases the private sector partner purchases a community's wildlife quota without actively involving the community in the management of the business (2007: 1–2).

Because of these problems, the adoption of the CBNRM Policy was "intended to guide and facilitate the strengthening of, and support to, existing and future CBNRM activities" (Ministry of Environment, Wildlife and Tourism, 2007: 2). The Policy aims to safeguard the interest of communities involved in NRM and to attract investment from natural resources based enterprises. The Policy would also provide regulation guidelines for community based organisation management practices to facilitate more equitable benefit sharing. The type of community–private sector partnerships pursued by this policy would afford the communities an opportunity to assume responsibility for business cooperation and adequately manage partnerships for the benefit of all community members. Based on these goals, the specific objectives of the CBNRM Policy can be summarised as:

- To provide guidance on CBNRM implementation on communal and state-owned land, except within such specifically excluded areas as officially gazette protected areas, forest reserves, national parks and game reserves
- To give communities incentives to engage in conservation activities leading to sustainable development and poverty reduction
- To establish an institutional, regulatory and participatory framework for the implementation of CBNRM; and

- To promote and ensure a responsible, accountable and transparent decision making process in CBNRM.

The Botswana National Ecotourism Strategy of 2002

The goal of the National Ecotourism Strategy (NES) is to create an environment in which *all* elements of tourism development planning and management facilitate, promote and *reward* adherence to the key 'principles' of ecotourism by all of those involved in the tourism industry. There are five main principles which have come to guide ecotourism development in Botswana:

- Minimising negative social, cultural and environmental impacts
- Maximising the involvement in, and the equitable distribution of economic benefits to, host communities and citizen entrepreneurs
- Maximising revenues for re-investment in conservation
- Educating both visitors and local people as to the importance of conserving natural and cultural resources
- Delivering a quality experience for tourists (without which tourists will not continue to visit, and so the benefits to conservation and development will not be sustained).

The NES promotes community-based tourism development in Botswana using any nature-based product that can be sold in the tourism market. The Okavango Delta is endowed with a variety of natural resources. As a result, if the various communities in the Okavango Delta are to benefit from tourism development in their local environment as the NES envisages, there is bound to be increased pressure in the use of natural resources in the wetland.

Ngamiland Tourism Development Plan of 2007

The vision of the Ngamiland Tourism Development Plan of 2007 is to strive for the development of a world-class nature-based tourism destination that is economically sustainable and optimises benefits to local communities and the nation within agreed limits of acceptable change. In its development of tourism in the Okavango Delta, the Ngamiland Tourism Development Plan is guided by the following ideals:

- Develop and promote Maun as the gateway to the Okavango Delta
- Develop a tourism gateway business centre at Maun Airport. This centre links the airport to the commercial area on the main road to north into a specifically identified and zoned tourism development node within Maun. A full range of tourism related economic, retail, service, travel, hospitality and restaurant facilities would be provided
- Broaden the permissible configuration of tourist lodges in the Okavango Delta. This concept challenges the current regulation limiting the number of tourist beds in a concession to 24 beds. It also promotes a range of different configurations of accommodation facilities permissible with a defined tourism devel-

opment node or nodes in concession areas. The number of permissible tourist facilities is dependent upon the defined, mapped tourism resources within the concession area

- Close part of Moremi Game Reserve to self-drive tourists. The idea behind this is to create opportunities for the provision of less sophisticated tourism products that may be provided by emerging, local tourism operators. These products include safari operating through the provision of safari vehicles and guides, game guides, non-permanent fly camps at certain locations within the reserve, hospitality products at north and south gates to Moremi Game Reserve
- Establish a development zone on the eastern side of Moremi Game Reserve that provides an appropriate range of experiences and facilities appropriate for the self-drive tourist market that provide optimal benefits to local host communities
- Establish a new national tourism asset in the form of Tsodilo National Park based on the San art and culture
- Establish a new tour route on the periphery of the Okavango Delta aimed at the overseas coach market-the Three Countries Triangular Tour Route.

The above tourism development ideals of the NgamilandTourism Development Plan indicate Government's intentions to develop tourism to its maximum in the Okavango Delta. Through this approach, no piece of land will be left unutilised in the Okavango Delta. If the Plan is to be effectively and successfully implemented, it will indeed increase tourism activities in the Okavango Delta.

Okavango Delta Management Plan

The Okavango Delta Management Plan (ODMP) is the Government of Botswana's initiative with the support of IUCN, German, Swedish and Danish governmental agencies such as German Development Service (DED), Swedish International Development Agency) and Danish Development Agencies (DANIDA) respectively in developing a management plan for the Okavango Delta. The overall objective of the ODMP is to carefully manage a well-functioning ecosystem of the Delta in a manner that ensures equitable and sustainable utilisation and provides benefits to local, national and international stakeholders.

The overall goal of the ODMP includes three strategic goals, each with several strategic objectives as outlined below:

- Strategic Goal 1: *To establish viable institutionalarrangements to support integrated resource managementin the Okavango Delta.*

  i) Strategic Objective 1.1: *To establish viable management institutions for the sustainable management of the Okavango Delta.*

  ii) Strategic Objective 1.2: *To improve the planning and regulatory framework for sustainable management of the Okavango Delta*

Tsodilo Hills, a unique San cultural site, is also a UNESCO World Heritage site currently being developed for tourism. Credit: Alyne Delaney

iii) Strategic Objective 1.3: *To raise public awareness, enhance knowledge and create a platform for informationexchange and learningabout the Okavango Delta.*

- Strategic Goal 2: *To ensure the long-term conservationof the Okavango Delta and the provision of existing ecosystemservices.*
  i) Strategic Objective 2.1: *To conserve the ecological character (biotic and abiotic functions) of the Okavango Delta, and the interactions between them;*
  ii) Strategic Objective 2.2: *To maintain or restore the wetlandhabitatsand ecosystems of Okavango Delta.*
- Strategic Goal 3: *To sustainably use the natural resources of the OD in an equitable way and support the livelihoodsof all stakeholders*:
  i) Strategic Objective 3.1: *To sustainably use the natural resources of the OD for the long term benefit of all stakeholders*;
  ii) Strategic objective 3.2: *To develop socio-economic opportunities to improve livelihoodsof the OD stakeholders.*

The development of ODMP will go a long way in ensuring integration of resource management and long-term conservation as well as provision of benefits of the present and future well-being of the people, through sustainable use of its natural resources.

Tourism resource uses

In the Okavango Delta, tourism activities are largely based on the commercial utilisation of wildlife resources. This is categorised into two broad groups, namely:
- Consumptive uses
- Non-consumptive uses.

Consumptive wildlife resource uses

Consumptive wildlife resource utilisation involves the process of wildlife off-take such as spot hunting or trophy hunting, game farming, live capture and export of live animal or translocation, and taxidermy and trophy processing. In the Okavango, because of sufficient wildlife resources, consumptive wildlife utilisation has taken the form of hunting tourism safaris. Safari hunting is done outside protected areas in Controlled Hunting Areas (CHAs) or in concession areas. Several hunting safari companies, mostly based outside Botswana but with offices in Maun, operate the hunting safaris in the region.

Safari hunters, mainly from industrialised western societies, visit the Delta for the purpose of sport hunting and for obtaining trophies. Scout Wilson Consultants (2001) note that safari hunting in the Delta starts in the United States where hunters from around the world attend the Safari Club International (SCI) convention in January every year. At this convention safari hunting companies sell hunts up to 2–3 years in advance. The report further notes that the majority of the hunters are Americans, followed by Spanish and Italians. Hunting quotas and hunting licences are provided by the DWNP to individuals, community organisations, and to safari hunting companies. These permits are provided for hunting during the hunting season from April to September each year. Safari hunting tourism has generated much revenue for local communities in Ngamiland and for Botswana as well as creating employment for the local people.

Non-consumptive wildlife resource use

Unlike consumptive wildlife utilisation, non-consumptive wildlife resource utilisation does not involve a wildlife off-take. The Okavango Delta is one of the major areas in Botswana where wildlife resources are important for scientific, educational and recreational purposes. Apart from hunting, wildlife use in the Okavango is mostly associated with photographic tourism. Photographic tourism involves activities such as photographic safaris, photographic camps and lodges, air charter operators, and the wildlife film industry. In addition to accommodation, camping, food and beverages, photographic tourism safaris in the delta offer game drives, boat safaris, and walking safaris. Some operators in the Delta also offer elephant and horse riding safaris. Photographic tourism is environmentally friendly and suits the interests of international animal rights groups (Mothoagae 1995). Among other things, this type of wildlife use is sensitive to the numbers of people involved. If the number is too high, this can disturb the animals and birds in an area and cease to be environmentally friendly.

Photographic tourism is especially important in Moremi Game Reserve, which offers what can be described as protected area safari tourism. The World Conservation Union (International Union of Conservation of Flora and Fauna-IUCN) defines a protected area as an area which contains exceptionally beautiful or unique landscape, fauna and flora of national and international importance. Wildlife-based tourism is the only type of land use allowed in these areas because it is assumed that tourism is compatible with the primary wildlife conservation goals of protected areas. The establishment of tourist facilities such as lodges and camps is done by private enterprise to promote tourist activitiesin Moremi Game Reserve. The primary management goals of the tourist use of protected area are to conserve the natural features and maintain biodiversity. Photographic tourism in the Okavango Delta is also carried out in what have come to be known as commercial photographic and community photographic areas.

- Types of Non-consumptive tourism activities
  Non-consumptive tourism activities in the Okavango Delta include the following:
  - Bird watching
  - Game viewing
  - Mokoro
  - Mobile safaris
  - Sex tourism
  - Fishing
  - Horse riding
  - Crafts
  - Walking trails
  - Hot air balloon.
- Kinds of tourists
  The kinds of tourists that visit the Okavango Delta can be categorised into the following:
  - Ecotourists
  - Business people
  - Study tourists
  - Family and friends tourists
  - Wildlife-based tourists (with different price preferences)
  - Safari hunting
  - Leisure tourists.

Segments of wildlife-based tourists

Wildlife-based tourists who visit the Okavango Delta can be sub-divided further based on price segments. These visitors can be categorised into non-consumptive

Table 7.1: Type of tourists and tourist activities in the Okavango Delta

Type of Tourist	Tourist Activities	Remarks
High cost or fixed	Game drives, night drives, walking trails, boating, bird watching, fishing, canoeing	High paying tourists, pay for visit as a package in countries of origin, stay in 2–3 camps for 2–3 nights (total 6–8 nights)
Mobile	Game drives (at times boating), bird watching, fishing, canoeing	Stay in private and public camps, spend 5–21 days, pay for visit as a package in country of origin, second highest paying tourists
Self-drive (independent, low cost)	Game drives, bird watching, fishing	Stay in public campsites, spend 8–10 days, drive own or rent vehicles, least spending tourists
Day Visitors	Game drives, boating, bird watching, canoeing	Visit the Delta in the morning and come back in the evening (they are largely conference and business delegates in Maun)
Safari hunters[1] (spot or trophy hunters)	Hunting (some combine it with photographic activities as in high cost tourists)	Some combine hunting with photographic activities hence classified as high cost tourists.

[1]The study did not focus on safari hunters hence the impact of safari hunting is not discussed in this chapter.

(e.g. photographic) and consumptive (e.g. trophy hunters) visitors. Photographic tourists include: independent (private), mobile, high cost (fixed lodge) and day visitors while consumptive tourists include safari hunters (trophy hunters). Safari hunters are not new to the Okavango Delta; they have been in the wetland since the 1850s (Tlou 1985). The tourism policy framework organised the safari hunting tourism sector through designated Controlled Hunting Areas.

The 'high cost' tourist

The most important type of tourist visiting the Okavango Delta is the 'high cost' tourist. A high cost tourist is defined as a person who chooses to stay at a permanent camp operated by private camp operators/owners (Borge *et al.* 1990). In the Okavango Delta, 'high cost' tourists mostly come from Europe, North America, and New Zealand/Australia These clients fly into Maun from either Johannesburg, Victoria, or Windhoek. From Maun they fly directly to permanent camps in the Okavango Delta. 'High cost' tourists rarely use accommodation facilities in Maun. They pay a high price for the package in advance and generally stay in high cost camps owned by the safari company from which they bought their package. On average, 'high cost' tourists pay USD 6,000 for a safari package in the Delta. They usually visit two or three camps owned and operated by the same company and spend 2 to 3 nights at each camp.

The mobile safari tourist

The second most important type of tourist visiting the Okavango Delta is the 'mobile safari' tourist. Mobile safari tourists stay in private, public or Hotel and Tourism Association of Botswana (HATAB) campsites. Like 'high cost' tourists, they fly into Maun from Johannesburg, Victoria Falls or Windhoek. From Maun they depart on a circuit that involves camping in the Okavango Delta. Mobile safaris take between 5 and 21 days with fees ranging from less than USD200 per night for camping (the least expensive option), USD200 – 500 per night (for modest priced packages) and over USD 5,000 per night for upmarket packages. Mobile tourists in the Okavango Delta include those who are on tours that encompass the wider east and southern African wildlife areas. Such trips are made on overland trucks and can take up to eight months, commencing in South Africa or Namibiaa nd ending in Kenya or Tanzania or vice versa.

The independent or 'low cost' tourist

The third group of tourists who visit the Okavango Delta are 'independent or low cost' tourists. Roberts *et al.* (1985) define the independent tourist as one who "travels independently, relying upon his own resources and not as part of an organised tour group". These travellers may drive independently, hire a guide, or charter a flight. Borge *et al.* (1990) state that their distinguishing feature is that they usually do not pay a package price before entering Botswana and they generally use public facilities such as campsites run by the Department of Wildlife National Parks in protected areas or privately owned campsites elsewhere. Independent tourists are also termed 'low cost' tourists because of their assumed low expenditure while in the Delta. Independent tourists in most cases visit Botswana from neighbouring countries such as South Africa, Zimbabwe, and Namibia. These tourists travel on private four-wheel drive vehicles such as Land Rovers or Land Cruisers with as many as five passengers in a single vehicle, and often in family groups. Independent tourists often visit Botswana for the first time to view the country's unique environment and to experience the unrestricted wildlife areas. At times independent and mobile tourists combine both types of travel in their journeys to northern Botswana.

Day visitors

Visitors or tourists who visit a particular place for pleasure for a period of less than 24 hours are referred to as 'day visitors'. The Department of Tourism (2001) defines a day visitor as a visitor who does not spend the night in a collective or private accommodation in the place visited. In the case of the Okavango Delta, day visitors mostly drive or fly from Maun in the morning into various parts of the Delta (either for *mekoro* safaris or game drives) during the day and come back in the evening. Day visitors do not stay in a safari camp or lodge in the Okavango but perhaps find lodging or campsites in Maun. As a result, when compared with the above three main categories of tourists the prices they pay are much lower.

Table 7.2: International tourist arrivals

Year	Arrival	Tourist Growth Rate (%)	Day Visitors	Total Visitors
1994	463,196		156,386	619,582
1995	521,041	12.5	115,440	636,481
1996	512,118	-1.7	143,845	655,963
1997	606,781	18.5	157,624	764,405
1998	749,544	23.5	190,413	939,957
1999	852,365	13.7	196,109	1,048,473
2000	969,291	13.7	201,974	1,171,264
2001	1,193,399	23.1	257,228	1,450,627
2002	1,273,814	6.7	211,355	1,485,169
2003	1,405,985	10.4	186,512	1,592,497
2004	1,522,847	8.3	203,870	1,726,717
2005	1,675,132	10.0	209,986	1,885,118
Average Annual Growth Rate 1994–2005	*13.7%*		*3.0%*	*11.8%*

Source: WTTC2007

Safari hunters

Safari hunters are also an important group of tourists in the Okavango Delta As already noted, safari hunters (i.e. spot or commercial hunters) mostly come from North America, Europe and New Zealand/Australia. A typical safari hunter in the Okavango Delta generates more revenue than a photographic tourist, so safari hunting contributes more money to Botswana's tourism industry than photographic tourism.

The fact that day and mobile tourists continue to visit indicates that the Tourism Policy has not necessarily discouraged other tourists in favour of those who stay in permanent accommodation, as it intended. As a result, additional numbers of tourists in the Okavango Delta can lead to crowding of facilities and pose threats to environmental sustainability of the wetland.

Tourist trend analysis

Tourism development in Botswana and the Okavango Delta in particular is conducted in the context of global and regional tourism trends. As a result, it is critical to analyse international and regional tourism trends in order to understand tourism trends in the Okavango Delta. Table 7.2 provides the number of international tourist arrivals by year from 1994 to 2004.

Table 7.3: Summary statistics for tourism demand in Botswana for 2005

International Tourist Arrivals	1,625,132
Average Annual growth rate: 1994–2005 (tourist arrivals)	13.7%
Average length of stay (nights)	5.5
International Tourist Expenditure(Pula)	2.4 billion
Average Spent per Person per Day (Pula)	274
Room Nights Sold	615,069
Bed Nights Sold	1,024,623
Room Occupancy Rate	41.5%
Visits to National Parks	237,258
Revenue at National Parks (Pula)	23.6 million

Source: WTTC, 2007

The number of tourists arriving in Botswana has grown significantly since 1994 as depicted in Table 7.2. Table 7.3 shows summary statistics for tourism demand in Botswana for 2005. According to the Botswana Travel and Tourism Satellite Account data, the rate of growth of tourist arrivals over the period from 1994 to 2005 averaged 13.7% per annum, exceeding the world average growth rate of around 4% in the same period. For 2007, Botswana Travel and Tourism was expected to grow 7.3%, and over the next ten years travel and tourism demand is forecast to average 5.0% growth per annum. This would still exceed expected worldwide growth of 4.4% per annum, and the Sub-Saharan African annual average forecast of 4.5%.

Figure 7.1 shows that at present tourism activities are concentrated largely in the northern part of the country, especially in the Chobe National Park and the Okavango Delta. According to data from the 2006 Department Wildlife and National Parks, of the P25.7 million park entry revenue generated, Chobe National Park contributed 60% (P15.5 million), while Moremi (OD) generated 27% (P6.94 million), the Pans 4% (P1.03 million) and the Southern Parks 9% (P2.31 million). The Okavango Delta and the Chobe National and the Pans together accounted for 95% of all park entries and 91% of all park revenues in 2006.

As this data illustrates, the growth of tourism could result in over-utilisation of these areas if proper steps are not taken to reduce pressure on the northern areas. Tourism product diversification is therefore necessary to ensure the distribution of resource utilisation in the country. The increase in numbers currently reflects increases in the dominant destinations of the Okavango Delta and the Chobe National Park.

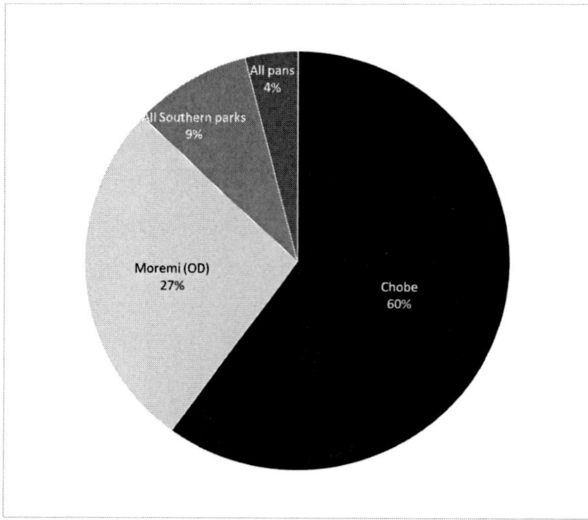

Figure 7.1: Overall revenue generated by national parks and game reserves in northern Botswana. Moremi Game Reserve located within the Okavango Delta
Source: Department of Wildlife and National Parks (2007)

Okavango Delta tourism

In their 1998 survey at border entry points in Botswana, the Department of Tourism (2001) estimated that 52,512 tourists visit the Okavango Delta annually. The number of visitors at Moremi Game Reserve has also been used to estimate annual tourist numbers in the Okavango Delta because it is located within the Delta. It is assumed that the majority of tourists who visit the Okavango also visit the Reserve. Table 7.4 shows data from DWNP annual reports which indicate that between 1995 and 2006, there was an increase of visitors in all the tourist segments at Moremi Game Reserve. For example, in 1971, only 4,500 private tourists visited Moremi Game Reserve (DWNP 1991), but this figure increased to 16,367 in 1999 and 12,585 in 2005. In 2000, 2001 and 2005, visitor numbers for private and mobile tourists went down because of heavy rains that made the roads impassable for driving, resulting in closure of the Reserve for several months (e.g. in 2000, the Reserve was closed for eight months).

The increase of tourist numbers at Moremi Game Reserve indicates that the Tourism Policy, which aimed at reducing tourist numbers in wilderness areas by charging higher prices, has not been successful. Ever increasing tourist numbers, if not well managed, threaten the ecological sustainability of the Okavango Delta as a natural system. Table 7.2 also shows that tourists that stay in permanent accommodation increased from 15,479 in 1995 to 24,103 in 2005. This means that sections of the Tourism Policy which aimed at increasing this tourist segment have been successful.

Table 7.4: Tourist numbers according to segment at Moremi Game Reserve

Year	Number of Tourists			
	Independent (private)	Mobile	High Cost (Fixed Lodge)	Totals
1995	8,025	12,570	15,479	36,074
1996	7,495	15,395	15,314	38,204
1997	11,298	17,487	14,202	42,987
1998	14,560	17,890	17,106	49,556
1999	16,367	14,835	15,487	46,707
2000	8141	8,521	14,173	30,835
2001	7035	7,687	16,351	31,073
2002	12,538	10,801	16,395	39,734
2003	10,659	8,650	18,069	37,378
2004	12,585	8,426	17,411	38,422
2005	9,431	7,979	24,103	41,513
2006	10,533	8,116	18,732	37,381

Source: DWNP Annual Reports (1995–2006)

Table 7.5: Tourist numbers according to country of residence, 1999–2006

Country	1999	2000	2001	2002	2003	2004	2005	2006
Citizens	3,761	2,787	3,353	3,728	4,63	2,849	2,344	1,359
Residents	3,456	2,595	5,5	3,24	2,679	2,12	1,568	709
South Africa	9,625	5,007	9,213	6,744	6,509	7,307	5,416	4,721
Other Africa	987	613	3,031	627	636	757	614	269
North America	7,461	6,154	8,331	6,82	6,147	6,435	10,197	129
South America	1,057	693	1,933	867	551	945	1,244	22
United Kingdom	3,762	2,426	4,968	3,184	3,267	3,406	4,034	292
Other Europe	14,122	8,993	19,625	11,978	12,982	13,056	14,356	2,862
Australia and New Zealand	1,761	1,148	2,55	872	1,351	1,204	1,319	164
Asia	347	109	875	258	318	235	241	5
Other	821	130	277	89	88	108	180	1
Total	47,16	30,835	31076	31,023	39,158	38,422	41,513	10,533

Source: DWNP Annual Reports (2000–2006)

International tourists and domestic tourists

The shift from casual campers to up-market tourists resulted in the majority of the tourists visiting the Okavango Delta coming from developed countries. Table 7.5 provides data from DWNP indicating that between 1999 and 2005 most visitors in the Okavango were from European countries, North America and South Africa.

Data in Table 7.5 also show that there were fewer citizen visitors in the Okavango Delta than those from other countries. This may be a result of high prices charged in tourism facilities in the wetland. The exclusion of citizens from visiting the Okavango due to high prices has the potential to create land use conflicts and negative perceptions by local people towards international tourism in the Okavango.

Tourism infrastructure development

There has been an increase in accommodation facilities in the Okavango Delta since the 1990s, when there were only 32 facilities (Mpotokwane 1990). This figure increased to 63 facilities in 2001, 73 in 2004 and 80 in 2009 (Mbaiwa 2009). However, this is likely to be a conservative figure as hunting camps used by safari hunters are not included in these figures.

In 2009, the Tawana Land Board Consultancy for Tourism Related Sites Identification in the Okavango Delta Ramsar Site found 80 tourism establishments in the Okavango Delta (excluding facilities in Maun). The Consultancy also shows the number of rooms and beds in these facilities (see Table 7.6).

Table 7.6: Number of beds and rooms in the Okavango Delta

Type of Facility	Rooms	Beds
Camp	407	810
Camp (allocated but not operational)	8	16
Guest House	31	51
Hotel	55	110
Hunting Camp	107	182
Lodge	201	374
Mobile	13	26
Photographic and Hunting Camp	6	8
Photographic/Hunting/Cultural Group	6	12
Photographic Camp	23	46
Grand Total	857	1635

An implication of the increase in the number of photographic and hunting camps in the Okavango Delta is that environmental pressure on the wetland is increasing. There is also likely to be competition and conflict over land use and resources in specific sites in the Okavango Delta. Studies (for example, Darkoh and Mbaiwa 2005) have shown that resource degradation occurs whenever there is competition and conflict over land use and resources. While there is an increase in tourism establishments in the Okavango Delta, the Okavango Jakotsha Community Trust (comprising of Jao, Etsha 1–13, Ikoga and Thaoge Villages) reports that the desiccation of the western part of the Okavango Delta is impacting negatively on their lodging business. The Trust argued that their lodge had not been operational for a few years partly because the Thaoge Channel (one of the Okavango

River channels) is dry and no longer attracts tourists. This underlines the point that flood levels in the Okavango Delta determine the level of tourism investment in the wetland.

Transport and communication systems

In the late 1980s, northern Botswana was almost inaccessible because there were virtually no tarred roads. The situation has gradually changed. During the National Development Plan 6 of 1986 to 1991, Government realised that tourism in north-western Botswana had the potential to contribute positively to the economy. At present, northern Botswana has over 1,622 kilometres of tarred road (see Table 7.7). Tarred roads link northern Botswana's main centres of Maun and Kasane with other eastern towns like Francistown and Gaborone. They also provide a link with the neighbouring countries of Zimbabwe and Zambia, and Namibia's Caprivi Strip. The road network in northern Botswana has improved access to the region and hence increased efficiency in the delivery of social services. The network is also an important aspect of tourism development in northern Botswana, especially in the Okavango and Chobe regions. As noted, tourism development in Botswana is linked to that of southern Africa since most tourists (i.e., self-drive, mobile and high-costs) who visit Botswana (e.g. the Okavango and Chobe) also visit Zimbabwe, South Africa, Namibia and Zambia.

Table 7.7: The road network in Northern Botswana, 1989–2001

Road Section	Length in Kilometres	Year Completion
Francistown – Kasane	500	1989
Nata – Gweta	115	1992
Maun-Shorobe	38	1992
Gweta – Maun	203	1993
Tsau – Gumare	103	1991
Maun – Sehitwa	99	1993
Gomare – Sepopa	95	1992
Sihitwa – Ghanzi	245	2001
Ghanzi – Mamono	208	1999
Shakawe – Mohembo	16	1995
Total	1622 +	

Source: Government of Botswana (1997) and Department of Roads (2002) (personal communication)

The development of the road network in northern Botswana is part of the government objective to improve the road system in the country. As shown in Table 7.8, there has been an expansion of the road network in Botswana from 1990–2002. During Botswana National Development Plan 8 (1997/98–2002/3), the development of roads comprised 58% of the investment allocation for the Ministry of Works Transport and Communications (Botswana Government 1997).

Table 7.8: Roads maintained by the central Government, 1994–1999 (in kilometers)

Year	Bitumen	Gravel	Earth/Sand	Total
1990	2565	960	4589	8114
1991	2831	932	4371	8134
1992	3603	2950	2148	8761
1993	3878	2739	2148	8761
1994	4177	2632	1957	8766
1995	4761	2087	1957	8805
1996	4729	2191	1841	8761
1997	4789	2131	1841	8761
1998	4969	2131	1841	8941
1999	5570	2688	1959	10217
2000	6082	2270	1865	10217
2001	6421	1937	1798	10156
2002(March)	6501	1857	1834	10192

Source: Botswana Government (1997), CSO (2002), Department of Roads (2002), Personal communication

Airports and airstrips

In addition to the tarred road network, in 1993 Maun Airport was improved through a new terminal and extended runway. Improvement of the airport attracted international flights connecting Maun, Johannesburg, Windhoek, and Victoria Falls. These links form the main air routes used by tourists who visit the Okavango Delta. In terms of daily aircraft operations, in 2001 it was recorded that Maun Airport was the second busiest airport in Africa after Johannesburg International Airport (Table 7.9). The local newspaper, *Ngami Times*, reported that, "Maun Airport is regarded as the second busiest international in Africa in terms of aircraft movements after the combined Johannesburg area (South Africa) airports of Johannesburg International, Lanseria, Rand and Grand Central" (2001:1). In 2000, Maun Airport had an average of 256 aircraft landing and departing each day during the peak tourist seasons of April to October. In the non-peak season of November to March, there were an average of 157 aircraft landings and departures each day.

Most of this air traffic is by small engine aircrafts that fly into the Delta either to transport tourists or carry supplies. In 2000, there were eight privately owned air companies operating a total of 44 small engine aircrafts in the Okavango Delta and using Maun Airport as the main base (Table 7.10).

Table 7.9: Aircraft novement at Maun Airport, 2000

Description	Total Movements	Percentage
Commercial		
International	2,846	7.0
Domestic	34,666	86.1
Non-Commercial	2,786	6.9
Totals	40,246	100.0

Source: DCA (2001)

Table 7.10: Private air companies and number of aircraft, 2000

Name of Air Company	Number of Airplanes	Percentage
Sefofane Air	15	34.1
Moremi Air	7	15.9
Delta Air	6	13.6
Northern Air	3	6.8
Swamp Air	2	4.5
Mack Air	7	15.9
Bushfree	2	4.5
Safari Air	7	15.9
Total	44	100.0

Source: Mbaiwa (2000)

In the Delta, there are about 23 privately owned airfields and seven government airstrips, excluding those operated by the Botswana Defense Force. In addition to the existing airport facilities in Maun and in the Okavango Delta, Government is in the process of expanding facilities at Maun Airport, especially the runway. This will allow Maun Airport to accommodate larger aircraft. The expansion of Maun Airport suggests an increase in volume of tourists, airport facilities, the number of small bush aircrafts and employment opportunities, especially for local people. While this is a positive economic development, care should be taken to prevent negative environmental impacts in the area especially in the Okavango Delta. This suggests that limits on the carrying capacity of both Maun and the Okavango Delta should be decided to determine the level of expansion and the volume of tourism that the wetland can accommodate.

The impact of tourism on rural livelihoods

The introduction of the CBNRM programme in the 1990s in Botswana has resulted in tourism benefits for some of the communities living in the Okavango

Table 7.11: Employment at Sankoyo village, Khwai and Mababe CBNRM Villages

	Sankoyo			Khwai			Mababe		
Year	Trust	JVP[1]	Total	Trust	JVP	Total	Trust	JVP	Total
1997	10	41	51	N/A	N/A	N/A	N/A	N/A	N/A
1998	11	51	62	N/A	N/A	N/A	N/A	N/A	N/A
1999	11	51	62	N/A	N/A	N/A	N/A	N/A	N/A
2000	11	51	62	5	N/A	5	15	37	52
2001	13	9	22	5	N/A	5	15	64	79
2002	35	56	91	8	N/A	8	16	64	80
2003	42	56	98	10	N/A	10	18	64	82
2004	48	56	104	15	40	55	41	18	59
2005	45	56	101	15	50	65	41	25	66
2006	46	56	102	19	55	74	41	25	66
2007	52	56	108	19	57	74	41	25	66

Source: Mbaiwa (2008)
[1] Joint Venture Partnership

Delta. While the level of benefit experienced differs from one community to another, these include:
- Participation of local communitiesin ecotourism
- Employment creation in remote areas
- Income generation for local communities
- Availability of meat at household level
- The provision of social services.

The three villages of Sankoyo, Khwai and Mababe are singled out to illustrate discussion of the impact of tourism development and CBNRM on rural livelihoods.

Employment opportunities in remote areas

Employment is one of the main benefits of CBNRM that has improved livelihoods in Sankoyo, Khwai and Mababe Villages. Employment is provided by both the safari companies that sub-lease community areas and by Trusts in villages. As shown in Table 7.11, at Sankoyo the number of people employed increased from 51 people in 1997, when the Trust started operating, to 105 in 2007. At Mababe, the number increased from 52 in 2000 to 66 people in 2007. At Khwai, the number increased from five in 2000 to 74 people in 2007.

At Mababe, the percentage of people employed is 22.8%, at Khwai it is 21% and at Sankoyo it is 28%. The number of people employed is significant considering the small populations of these villages – 300 in Mababe, 290 in Khwai and 372 in Sankoyo. Illustrating the importance of this employment, the Mababe Trust manager remarked, "Go to Mababe right now and you will find zero unemployment. You will only find old people and children in the village. All the young and strong people are out in camps working." CBNRM is the most important economic activity providing employment opportunities in the three villages. Most people employed from the three villages are semi-skilled (e.g., cooks cleaners, storekeepers and escort guides). The creation of jobs in CBNRM projects is important as a poverty alleviation strategy. It should be noted that Ngamiland District in which the Okavango Delta is located does not have any industrial or manufacturing enterprises to provide employment opportunities as is the case with other centres such as Francistown or Gaborone. As such tourism is the main economic activity that employs a significant number of people, more than any other sector in the District. Every economy aims at full employment for its labour force, and this is also the case for the rural economies of Sankoyo, Khwai and Mababe.

Financial benefits from tourism development

Financial benefits are some of the major economic benefits that villages derive from tourism development in the Okavango Delta. Tourism revenue that accrues to communities is largely from the following activities: sub-leasing of the hunting area; sale of wildlife quota (i.e. wildlife quota fees for game animals hunted); meat sales; tourism enterprises (e.g. lodge and campsite); and, camping fees and vehicle hires. Income from tourism development accrues to individuals, households and the community at large when it is finally distributed. Table 7.12 shows the financial benefits that accrued to Sankoyo, Khwai and Mababe respectively from the time the projects started operating to 2007.

Data shown in Table 7.12 show that land rentals and quotas increased in each village over time. For example, at Sankoyo land rentals increased from P285,750 in 1997 to P483,250 in 2006 and game quota fees increased from P60,928 in 1998 to P1,198,700 in 2006. Between 2004 and 2006, game quota fees were the largest source of revenue for each village, accounting for almost half of the revenue generated by each Trust. Income generation from tourism is important because it is used by communities to complement and sustain their livelihoods. Some of the income which the three communities derive from tourism subsequently ends up in the households in the form of dividends. For example, between 1996 and 2001, each household at Sankoyo Village was paid P200, and this sum increased to P250 in 2002, P300 in 2003 and P500 between 2004 and 2007. The distribution of income to the various households is an important aspect of improving rural livelihoods.

Availability of meat at household level

Households at Sankoyo, Mababe and Khwai noted that game meat is one of the benefits they derive from tourism. Joint venture agreements with safari operators mean that international clients of these companies only take the animal heads and

Table 7.12: Revenue (BWP) generated by Sankoyo, Khwai and Mababe

Year	Land Rental	Quota	Others[1]	Total
Sankoyo Tswaragano Management Trust				
1997	285,750	0	12,665	298,415
1998	116,666	60,928	38,826	216,420
1999	151,667	33,470	76,151	261,288
2000	166,833	49,090	148,940	215,923
2001	57,047	55,600	114,801	227,448
2002	492,000	872,550	131,844	1,496,394
2003	466,509	965,772	370,352	1,802,633
2004	562,655	1,096,377	75,634	1,734,666
2005	455,000	1,060,400	612,012	2,127,412
2006	483,250	1,198,700	639,116	2,321,066
2007	613,360	1,272,600	621,537	2,507,497
Khwai Development Trust				
2000	1,057,247	0	72,536	1,129,783
2001	585,220	0	248,305	833,525
2002	1,211,533	0	36,738	1,214,567
2003	348,778	0	97,480	446,258
2004	110,000	857,085	283,482	1,250,567
2005	115,500	1,043,707	405,247	1,564,454
2006	121,275	1,248,500	1,248,500	1,691,723
2007	127,339	1,217,187	1,082,146	2,426,667
Mababe Zokotsama Trust				
2000	60,000	550,000	77,000	687,000
2001	69,000	632,500	127,233	828,733
2002	79,350	702,606	85,961	867,917
2003	91,205	807,996	98,854	1,121,427
2004	104,940	929,196	149,159	1,183,295
2005	120,681	1,068,575	130,739	1,319,995
2006	120,000	1,202,183	13,500	1,335,683
2007	130,000	1,202,183	29,950	1,362,133

Source: Mbaiwa (2008)
[1]Camp rental fees, community development fund, meat sales and vehicle hire.

skins (called the trophies) and fillets. The rest of the carcass is taken to the respective village where it is either sold or distributed to the community depending on the type of animal killed. For instance, animals such as elephant, warthog and

Table 7.13: Social services funded with CBNRM income, 2007

– Assistance with funerals costs (P200 to P3, 000 per household).
– Support for local sport activities (P5, 000 to P50, 000 per village).
– Scholarships (P7, 000 to P35, 000 per village).
– Household dividends (P200 to P500 per village).
– Services and houses for elderly people (150 to P300 per month per person).
– Assistance for orphans (P40, 000 per Trust).
– Assistance for disabled people (P15, 000 per village).
– Provision of communication tools such as radios
– Transport services particularly in the use of vehicles.
– Installation of water stand pipes in households

Source: Arntzen et al. 2003; Schuster 2007; Mbaiwa 2008

zebra, which are not normally eaten by most residents, are distributed free. However, meat such as that of impala and kudu and other animals such as buffalo, which are eaten by most people, are sold by the Trust at a minimal price (compared to beef sold in other parts of the country) and the money becomes part of the overall Trust revenue. Some community members pointed out that the distribution of meat benefits is important to them since there is no butchery in the village, indicating how important this source of meat is. It also allows the villagers to benefit directly from wildlife resources in their local environment. This has the potential to reduce illegal hunting which was higher in these villages before the adoption of CBNRM.

The provision of social services

The people of Khwai, Mababe and Sankoyo agree in *kgotla* meetings on how to use money generated from tourism to benefit their villages. These communities have been able to agree that some of the revenue generated from CBNRM projects be used to fund social services and related community development projects. Table 7.13 shows the social services and amount used to fund each in the three villages in 2007.

To elaborate on the information provided in Table 7.13, collective decision-making in the three communities has led to the payment of funeral expenses for community members to the tune of P3,000 if an adult dies and P1,000 if a minor dies. Community trust vehicles would also be used to collect firewood for preparing meals for mourners and for transporting them to and from Maun. Apparently, residents from all three villages bury their dead in Maun although there seemed to be no specific reason why they do not bury their dead in their respective villages. Community vehicles are also used to transport the corpse and are a form of transportation by members from one village to the other. Community members can also hire these vehicles to transport goods. This is yet another important contribution of tourism to community and rural development in remote parts of the Okavango Delta Sankoyo Mababe and Khwai are located in remote areas with little or not access to public transport. The availability of transportation through Trust vehicles has therefore increased accessibility for residents of these once remote areas to other big regional centres in other parts of the country like Maun.

Before tourism through CBNRM, water reticulation was a problem at the three villages. To address this problem, the communities used income from their CBNRM activities to fund water reticulation to each household. For example, the CBNRM project at Sankoyo funded the provision of water to 56 of the 76 households in the village. Mababe funded water to 30 out of the 54 households in the village. At the time of the study, Khwai was still digging a trench from the Khwai River to provide water to their village. The other social service funded by CBNRM in the three villages was housing: in Sankoyo, CBNRM funds paid for the construction of 7 houses for the poor in 2007, at Khwai 18 houses were built, while at Mababe 10 houses were built for elderly and poor members of the community.

Furthermore, in each of the three villages, CBNRM paid a monthly allowance of P200 to orphans and P500 to the elderly twice a year. Sankoyo sponsored 14 students to study for catering, professional guiding, bookkeeping, and computer studies. Mababe sponsored 20 students and Khwai sponsored 30 students by 2007, with a total sponsorship of P250,000 to study tourism-related courses like those for students from Sankoyo. For Khwai, all the villagers were insured with insurance companies for funeral benefits. CBNRM in the three villages has provided livelihoods and social services for community members, thus transforming the communities at Sankoyo, Mababeand Khwai from dependency on hand-outs from the Botswana Government and donor agencies in Europe and America into productive communities that are moving towards achieving sustainable livelihoods.

Conclusion

It is evident from the above account that tourism in the Okavango Delta has grown tremendously in the past few decades and benefits the economy and some communities in the District. The various policies and development plans aimed at promoting tourism in the Okavango have yielded notable progress in tourism development in the wetland. As a fragile ecosystem, much effort has been take to ensure that development does not harm the environment in the Delta. The various government policies advocate for equitable sustainable use of the resource, and have put in place baseline data and clear monitoring and evaluation mechanisms to ensure that development of the Delta, complements the set conservation objectives.

It is not clear however how the economic and environmental benefits outlined above will be translated into social equity. Despite the benefits to some communities as shown above there are questions about the involvement of local people in Ngamiland generally and their ability to take advantage of such benefits. Furthermore it is questionable how involved they are in the development of the policies and strategies and whether these enable community access to resources. It is important to note that institutional policies and modern state practices have played a key role in regulating natural resources in rural areas, and are notorious for limiting local access. Botswana and the Okavango Delta are no exception as shown by a proliferation of policies and other regulatory instruments for tourism development. All too often policies, institutions and strategies deliberately displace people living within proposed preserves, and curtail or ban their customary rightsof access to natural resources in these areas, which invariably creates antagonisms between displaced or disposed communities and protected park or wildlife managers (Mbaiwa *et al.* 2008). The challenge for natural resources governance in developing countries is to work out institutional frameworks that place rural subsistence communities and access to natural resources at the centre of natural resource use planning in order to achieve sustainability in both livelihoods, resource use and social relations.

References

Arntzen, J., Molokomme, K., Tshosa, O., Moleele, N., Mazambani, D., and Terry, B. (2003). *Review of CBNRMin Botswana,* Centre for Applied Research, Gaborone.

Borge, L., Nelson, W.C., Leitch, J.A. and Leistritcz, F.L. (1990). *Economic Impacts of Wildlife-based Tourism in Northern Botswana* Department of Agricultural Economics, North Dakota State University, Fargo, North Dakota.

Darkoh, M.B.K. and Mbaiwa, J.E. (2005). *Natural Resource Utilization and Land UseConflicts in the Okavango Delta, Botswana* Department of Environmental Science and Harry Oppenheimer Okavango Research Centre, University of Botswana, 112 pp.

Department of Tourism, DOT. (2001). A Statistical Review of Tourism 1998–2000, Department of Tourism, Gaborone.

Department of Wildlife and National Parks (DWNP). (1991). *Moremi Game ReserveManagement Plan, Draft Report,* Department of Wildlife and National Parks, Gaborone.

Government of Botswana. (2003). *Botswana 2003 Second Generation HIV/AIDS Surveillance Report*, National AIDS Coordinating Agency 2003.

Government of Botswana (1986). *Wildlife Conservation PolicyGovernment Paper No. 1 of 1986,* Government Printer, Gaborone.

Government of Botswana (1990). *The Tourism Policy. Government Paper No. 2 of 1990,* Government Printer, Gaborone.

Government of Botswana (1997). *National Development PlanVIII 1997–2003,* Government Printer, Gaborone.

Mbaiwa, J.E. (2002). "The Socio-Economic and Environmental Impacts of Tourism Development in the Okavango Delta, Botswana: A Baseline Study", Harry Oppenheimer Okavango ResearchCentre, Maun, Botswana, July 2002, 144pp.

Mbaiwa, J.E. (2008). "Tourism Development, Rural Livelihoods and Conservation in the Okavango Delta, Botswana", PhD Dissertation, Department of Recreation, Park and Tourism Sciences, Texas A&M University, 186pp.

Mpotokwane, M. (1990). "Tourism and Environmental Issues in Botswana."*In*: Pfotenhauer (ed.), *Tourism in Botswana,* Proceedings of a Symposium held in Gaborone, 15–19th October 1990 (pp. 146–157), Botswana.

Schuster, B. (2007). "Proceedings of the 4th National CBNRM Conference in Botswana and the CBNRM Status Report", 20–23rd of November 2006, IUCN Botswana, Gaborone, Botswana.

Scout Wilson Consultants. (2001). "Integrated Programme for the Eradication of Tsetse and Tyrpanosomiasis from Ngamiland: Environmental Impact Assessment, Draft Final Report", Scout Wilson Resource Consultants, Edinburgh.

Roberts, M., Ingram, B. and Fowkes, J.S. (1985). "A profile of the Independent Tourist Visiting Botswana in 1985." Kalahari Conservation Society, Gaborone.

Tlou, T. (1985). *History of Ngamiland: 1750–1906: The Formation of an African State,* Gaborone: Macmillan Publishing Company.

World Travel and Tourism Council, WTTC (2007). *Botswana: The Impact of Travel and Tourism on Jobs and the Economy,* London: World Travel and Tourism Council.

8 Water Resource Management and Irrigation Development in the Okavango Delta, Botswana

Wellington R.L. MASAMBA and Moseki Ronald MOTSHOLAPHEKO

Abstract

The use of water resources, particularly for irrigation and arable production, has been identified as a plausible planning and policy approach for improving food security and reducing poverty in many developing countries. In Botswana, water resources management has recently been assessed with a view to adopting an integrated water resources management (IWRM) approach. This chapter assesses water resources management with a particular focus on the extent of irrigation and its possible contribution to food security in the Okavango Delta and Botswana in general. The chapter further explores implications for planning and policy development. Data were obtained from secondary sources including review of reports on water management, and from primary sources including interviews with public officers in some key departments involved in water resources management and irrigation. Indications are that IWRM is steadily being adopted, but there are institutional and resource challenges that limit full implementation. Botswana is a semi-arid country and due to limited surface water resources, the ephemeral nature of most river catchments, unreliable rainfall, generally poor soils and technological challenges, arable farming through irrigation is limited. This is the case even in areas with high potential for irrigation such as the Okavango Delta. Due to a lack of irrigation policy in Botswana, there is limited guidance on water use for irrigation in the Okavango Delta. Even with the implementation of widely consultative plans such as the Okavango Delta Management Plan, issues of water use for improving food security and reducing poverty remain indeterminate. There is a need for irrigation policy to guide water use particularly in trans-boundary water resources such as the Okavango Delta.

Key words: water resources management, irrigation, policy and planning

Introduction

Globally freshwater resources are limited and under stress due to increasing and often competing human demands. As a result, water resources management, which entails optimising water use and minimising the environmental impacts thereof, has over time shifted in emphasis from supply side to demand management, then to integrated water resources management. The latter, derived from the ecosystem approach, is currently being promoted with an emphasis on natural resources management (NRM) and sustainable rural livelihoods. Water resources management has also been seen as an important and all-embracing area, including sanitation and hygiene. Many countries of the world, particularly in the global South, have been plagued with challenges of access especially among the majority poor members of the society.

In Botswana, the management of water resources has been centrally structured with the Government playing all the key roles of water development, regulation, planning and policy-making, and service provision, with some level of decentralisation to various sectoral ministries (Snowy Mountain Engineering Consultants [SMEC] 1991; Swatuk and Rahm 2004; Rahm *et al.* 2006; Swatuk and Kgomotso 2007). Furthermore, water resources management in Botswana is based on administrative regions rather than river catchment or river basins. Several challenges in the water sector have been identified ˉ including lack of coordination, inefficiencies and uneven level of services for water supply and sanitation provision throughout the country (World Bank 2009; Swatuk and Rahm 2004). Following the recommendations of the National Water Master Plan Review of 2006, some reforms to the water sector have been initiated; the Ministry of Minerals, Energy and Water Resources (MMEWR) is responsible for national water policy and development whereas the Water Utilities Corporation is responsible for water and sanitation service delivery throughout the country (World Bank 2009). However, the roll-out of service provision to urban villages and rural settlements by the Water Utilities Corporation has been slow and riddled with implementation challenges, including human resource shortages, inadequate maintenance and a general failure to meet demand for such services.

Irrigation has been identified as an important tool for supporting food security and alleviating poverty, particularly in arid and dry sub-humid regions of the world. In Botswana, the Ministry of Agriculture has emphasised the need for irrigation to develop the agricultural sector and improve its overall economic contribution. Irrigation is required in most areas as a result of the poor performance of rain-fed agriculture attributed to, among other reasons, inadequate and unreliable rainfall (SMEC 2006a). Annual rainfall ranges from 250 mm in the extreme south-west to 650 mm in the extreme north-east. Most of the rainfall occurs in summer when moisture loss to the atmosphere is high and this tends to limit the amount of moisture available for crops to grow and mature.

Irrigation development in Botswana focuses mainly on expanding horticultural production to meet a large percentage of domestic demand and to generate employment. With the implementation of the National Master Plan for the Arable Agriculture and Dairy Development (NAMPAADD), a total of 5,200–5,400 hectares was planned for irrigation by the year 2012 (from a total of 1,800 hectares in 2004); with a proposed 3,600 hectares to be irrigated using treated wastewater. The total amount of irrigation water required to meet these targets was estimated at 51,000 Ml/year to 53,000 Ml/year (SMEC 2006b).

Flood recession farming which is the traditional form of irrigation remains popular in the OD while modern technical irrigation remains underdeveloped. Photo Credit: Lin Cassidy

The contribution of the agricultural sector to the economy of Botswana in terms of Gross Domestic Product (GDP) is fairly small compared to other sectors such as mining, manufacturing, trade and tourism. According to the Ministry of Finance and Development Planning (2009), the sector's contribution to GDP has declined in importance from a high of 42.7% at independence in 1966 to a low of 2.1% in 2011 (Table 8.1). This decline has been attributed to the growth in other sectors particularly mining, commercial and administrative services, and tourism. The decline has also been marked by a general intergenerational shift from peasant agriculture to urban-based formal employment of mostly the youth. This trend may be reversed through technological advances and commercialisation of agricultural activities through appropriate programmes. The Government of Botswana continues to promote the agricultural sector with a view to improving food security and employment creation, especially for people in rural areas where agriculture accounts for 80% of employment. Some of the programmes introduced include the Integrated Support Programme for Arable Agricultural Development (ISPAAD), Livestock Management Infrastructure Development Programme (LIMID) and NAMPAADD. The latter, as indicated earlier in this chapter, is more relevant to irrigation projects as it focuses on supporting technological improvements and commercialisation of agriculture.

Table 8.1: Percentage (%) contribution of various economic activities to GDP by year and total GDP (constant 1993/94 prices) in million Pula.

Economic activity	Year and percentage (%) contribution to GDP															
	1966	1975/76	1985/86	1999	2000	2001	2002	2003	2004	2005	2006	2007	2008	2009	2010	2011
Agriculture	42.7	20.7	5.6	2.9	2.6	2.4	2.0	2.2	1.9	1.8	1.7	1.7	1.8	2.1	2.0	2.1
Mining and quarrying		17.5	48.9	38.1	40.9	41.1	42.0	42.2	40.5	40.6	41.7	39.0	36.5	30.4	30.3	28.4
Manufacturing	5.7	7.6	3.9	4.1	3.8	3.8	3.4	3.3	3.3	3.5	3.4	3.8	3.8	3.8	3.8	4.1
Water and electricity	0.6	2.3	2.0	2.1	2.2	2.2	2.2	2.2	2.2	2.2	2.2	2.2	2.3	2.4	2.4	2.4
Construction	7.8	12.8	4.6	6.0	5.2	4.7	5.8	4.7	4.9	4.5	4.4	4.8	4.5	5.1	5.5	6.6
Trade, hotels, restaurants	7.8	12.8	4.6	10.1	9.4	9.7	9.5	10.0	9.7	9.1	9.5	10.0	10.4	11.2	11.5	11.5
Transport	4.3	1.1	2.5	3.6	3.4	3.4	3.3	3.0	3.0	3.3	3.5	3.7	4.1	4.8	4.8	4.9
Banks, insurances, Business Services	20.1	4.7	6.4	10.2	10.0	10.3	10.0	9.5	9.3	10.0	10.2	10.5	11.7	12.8	12.5	12.4
General Government	9.8	14.6	12.8	14.7	14.8	15.1	15.5	15.8	16.1	16.6	15.6	15.5	16.3	17.7	16.6	16.6
Social and personal services		2.8	2.5	3.8	3.9	3.9	3.5	3.5	3.7	4.0	3.9	3.9	4.0	4.7	4.5	4.5
Total GDP (million BWP)	908.4	2222	5353.9	-	-	15165.6	-	-	-	21481.4	23639.6	24776.4	25504.7	24274.2	25976.7	27289.4

Source: Ministry of Finance and Development Planning (2003), and Central Statistics Office (2012). Note: Total percentage will not add to 100 because adjustment items, such as subsidies, taxes on imports and production, were excluded.

The Ministry of Agriculture (MoA) engaged TAHAL Consulting Engineers in 1998 to prepare NAMPAADD. The objective of this Master Plan was to improve the performance of the agricultural sector and to ensure sustainable use of the country's resources. NAMPAADD's planning horizon was 10 years from 2002 to 2012. It considered two scenarios for agricultural development: a socially oriented scenario where rural employment and household food security take precedence over economic efficiency; and an economically oriented scenario where competitiveness and the most economically efficient of resources assume precedence. The potential for employment was envisaged at 50% under the NAMPAADD socially-oriented scenario (TAHAL Consulting Engineers, 2000). The challenges faced by farmers before the start of NAMPAADD included a lack of draught power, deficient sector-wide strategies, untargeted support services, and disruptions from some policy interventions (Ministry of Agriculture n.d.). Six years since implementation, NAMPAADD was credited with the transfer of skills among farmers and extension workers, but it was limited by production in the intended productive areas. The slow pace of adoption of innovations such as weed control and other measures by farmers, mainly in the traditional arable farming sector, were identified as some of the challenges leading to low production.

This chapter reviews the management of water resources in Botswana in order to determine the general extent of irrigation, its management and general contribution to food security, and implications for planning in the Okavango Delta. This can improve knowledge on water resources management with regard to irrigation in Botswana, and the Okavango River Basin in general. The chapter contributes to ongoing efforts to determine sustainably viable paths for water resources management in the Okavango Delta and for wider application in the management of natural resources in the Okavango River Basin. The observations made on conditions in the Okavango Delta may be useful for planners, academics, and water practitioners involved in decision-making surrounding water use and development in similar conditions in other parts of the world. The chapter draws on the work undertaken as part of a trans-boundary diagnostic assessment (TDA), which formed part of the Environmental Protection and Sustainable Management of the Okavango River Basin (EPSMO) Project of The Permanent Okavango River Commission (OKACOM).

Study area

The general study area is the Okavango River Basin which emanates from the highlands of south-central Angola in the Cuito and Cubango river catchments. The Okavango River Basin remains one of the least human impacted basins on the African continent (The Permanent Okavango River Commission [OKACOM] 2013). At a broad scale, water resources management in the Okavango River Basin is guided by the Revised Southern African Development Community (SADC)

Protocol on Shared Watercourses of 2000 and related pieces of agreed principles for its implementation such as the SADC Regional Water Policy (RWP) of 2005 and the SADC Regional Water Strategy (RWS) of 2006. The protocol forms the legal instrument outlining the general obligation for member states to cooperate and consult each other on the development and use of water resources in all internationally shared river courses. The policy provides the context and intent for water resources management in the SADC region and the strategy outlines guidelines for implementation of the RWP through watercourse institutions (SADC 2006). At a global scale these instruments are linked to the Dublin Principles which underpin the IWRM approach in the water sector. Water use for irrigation in a shared watercourse such as the Okavango River Basin can best be understood within this broad regional and global context.

Mounting socio-economic pressures emanating from the need to develop and use water resources in the riparian countries of Angola, Botswana and Namibia has been observed to threaten the generally pristine environmental conditions in the Okavango River Basin. Therefore OKACOM solicited funds from the riparian states and the Global Environment Facility (GEF), and initiated the Environmental Protection and Sustainable Management of the Okavango River Basin (EPSMO) Project.

The long-term objective of the EPSMO Project is to achieve global environmental benefits through concerted management of the naturally integrated land and water resources of the Okavango River Basin. The specific objectives of the project are to:

- Enhance the depth, accuracy, and accessibility of the existing knowledge base of basin characteristics and conditions and identify the principal threats to the trans-boundary water resources of the Okavango River Basin through a Trans-boundary Diagnostic Analysis (TDA);
- Develop and implement, through a structured process, a sustainable and cost-effective program of policy, legal and institutional reforms and investments to mitigate the identified threats to the basin's linked land and water systems through the Strategic Action Program (SAP); and,
- Assist the three riparian nations (Angola, Botswana and Namibia) in their efforts to improve their capacity to collectively manage the basin. (OKACOM, 2011)

The SAP, which has been finalised, includes baseline and additional actions to address priority trans-boundary issues and serves as a monitoring and evaluation tool for implementation. It also recommends the development and testing of a set of institutional mechanisms and implementation methodologies, including pilot demonstrations that explicitly link regional, national and local initiatives in land and water management (OKACOM 2011). Additionally, it involves preparation of a basin-wide framework in which trans-boundary priorities can be addressed and project interventions monitored.

The TDA informed and guided the development of the SAP and acted as a platform where trans-boundary externalities could be examined and resolved. Furthermore, the TDA underpinned the SAP design, informing the monitoring and reporting criteria for SAP implementation. Most importantly, the process of completing the TDA informed policies and initiatives to be launched in preparation for SAP implementation. The TDA also assessed the extent of irrigation, its importance to the basin communities as well as plans for its expansion in the Okavango River Basin.

In this chapter, the specific study area is the Okavango Delta, one of the few surface water sources found in Ngamiland District of north-western Botswana. This District has a population of 158,104 people, with an annual growth rate of 2.1% per annum observed in the 2001 to 2011 inter-censal decade (Central Statistics Office, 2011). A large proportion of this population resides in rural settlements around the Delta. The level of poverty (people living below the poverty datum line) in Ngamiland District is 47.3% (Ngamiland West), which is higher than in most districts in Botswana (Statistics Botswana 2011). Most members of communities in the Ngamiland District practice livestock and arable farming. Due to poorer soil fertility in areas around the Okavango Delta compared to the floodplains, and the generally low rainfall, flood recession cultivation is the most favoured type of arable farming among some communities in the Delta. It is often accompanied by limited irrigation undertaken in small horticultural gardens. With the continued growth in district population, and a related rise in demand for arable produce, the need for irrigation may increase in future.

Since the declaration of the Okavango Delta as a Ramsar site in 1997, water resources management is undertaken within the principles of the Ramsar Convention. The convention requires contracting parties to adopt an ecosystems approach in the management of resources in wetlands. As a result, the Government of Botswana, with assistance from the Ramsar Secretariat, the Danish Development Agency (DANIDA), German Development Service (DED), the Swedish International Development Agency (SIDA) and the World Conservation Union (IUCN), developed the Okavango Delta Management Plan (ODMP) of 2008. The plan aims to integrate natural resources management for the Okavango Delta and ensure long-term conservation in order to provide benefits for present and future well-being of the people (Department of Environmental Affairs 2008). Therefore the plan serves as the main guiding tool for development and management of water and other resources in the Okavango Delta. The plan also supports regional engagement with other riparian states regarding the integrated management of terrestrial and aquatic resources in the Okavango River Basin. The contribution of this Plan to irrigation management in the Okavango Delta is discussed in Section 5 of this chapter.

Methods

Information was collected from a knowledge-gathering workshop, discussions with Government Officers and a literature review.

Knowledge-gathering workshop

A knowledge-gathering workshop was conducted on 24–25 February 2009. The participants were asked beforehand to prepare information on irrigation, and this information was obtained during the workshop. In addition, participants also provided input on their expectations and what should be included in the report. A feedback workshop was conducted from 15 to 16 July 2009, with additional input provided by the workshop participants.

Discussions with Government Officers

Additional information was obtained during visits to offices and interviews with officers as well as from recent and archival records (Agricultural Research, Maun, Regional Agricultural Office, Maun, NAMPPAARD offices in Gaborone).

Literature

Secondary data were sourced from published and unpublished reports obtained from Government offices and libraries. Most of the information in this report is derived from SMEC (2006), the National Water Master Plan Review (NWMPR), Volume 8, Agriculture.

Irrigation in Botswana

Irrigation in Botswana is mainly for high value horticultural and citrus production. Whereas small-scale farmers dominate other forms of agriculture, large-scale farmers dominate irrigated horticultural production. Irrigation is predominantly carried out in the Tuli Block area of eastern Botswana, which has high yielding wells and good soils along the Limpopo River. The farmers in this area also have standby pumps to extract water from the river during the months when there are flows. There are many small-scale farmers, most of whom have one hectare or less, chiefly because of low borehole yields. This denies these growers the benefits of economies of scale. The potential for commercial horticultural production is largely underutilised, even within the limitations of water supply. In regards to irrigated agriculture, especially horticultural production, the objective of the Government is to increase domestic production to meet 70% of domestic horticultural demand. This is motivated by the need for food security and employment generation (NAMPAADD Government White Paper No. 1 of 2002).

The history of planned water usage in the Okavango Delta has been reviewed in some studies (Randall 1957; Scudder *et al.* 1993). The proposals in Randall (1957) focused on draining the Okavango Delta for crop production. Pike (1970)

Table 8.2: Total area developed for irrigation as of December 2004.

Region	Area (ha)
Western	25.36
Southern	240.85
Gaborone	288.36
Central	593.90
Francistown	273.53
Northwest	339.00
TOTAL	1760.97

Source: Regional Project Inventories (2004) quoted in SMEC (2006)

evaluated the Okavango as a primary water resource, and irrigation zoning around the Nhabe River in Ngamiland was carried out for approximately 1,100 ha of land (van der Sluis 1990). Prior to this, the Southern Okavango Development project proposed the potential development of 10,000 ha for irrigated agriculture along the Delta's southern fringe (SMEC 1987). This project was later abandoned by the Botswana Government following local and international pressure resulting from environmental concerns. Currently no major irrigation is planned for the Okavango Delta region until international agreements for the use of the waters of the Okavango are signed.

Irrigation development is guided by several agricultural initiatives including NAMPAAD, and it remains fragmented between different sectoral ministries and departments. In the absence of a policy on irrigation, the only guiding document for managing irrigation development is the Irrigation Policy and Strategy, which has been in draft form since 2006. Irrigation development in Botswana is housed at the Department of Crop Production in the Ministry of Agriculture. As a subsection of this department the Irrigation Unit remains poorly resourced in terms of human, financial and other capabilities. As a result there is general lack of support for farmers engaged in irrigation projects and poor enforcement of regulations on user rights, and waste water use (Department of Crop Production 2006)

Present irrigation and plans for expansion

The majority of farms using irrigation grow horticultural products while a few small-scale farms grow field and fodder crops. Estimates show that a total of 1,800 ha was cultivated under irrigation to produce vegetables and fruits in 2004 (Regional Project inventories 2004 quoted in SMEC 2006), and this amount had increased by 67% to 3,000 ha in 2010 (Ministry of Finance and Development Planning 2010). Table 8.2 provides a summary of the irrigated areas in 2004 by region. The Central District has the highest level of irrigation; these are mostly commercial farms located along the Limpopo in the Tuli Block area. Small-scale irrigation

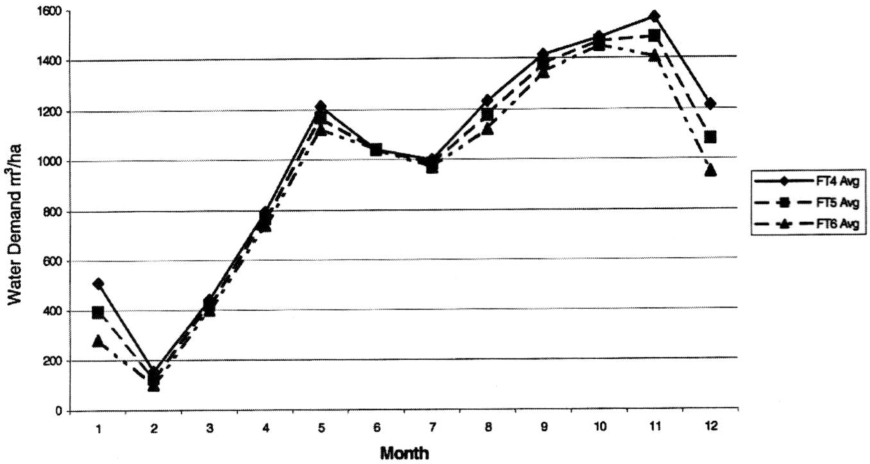

Figure 8.1: Average monthly water requirements for NAMPAADD Scenarios. FT4-Irrigation developing, FT5-irrigation established and FT6-irrigation advanced. Source: NAMPAADD quoted in SMEC 2006

projects have been developed in other parts of the country, including along the Tati River in the vicinity of Francistown and in Gaborone and the Southern Regions.

Water demand

The extent of irrigation development in Botswana is constrained by the seasonal availability of water. In the absence of storage and relocation facilities, the maximum area that can be cultivated is the area that can be adequately supplied with water during the period of highest demand. Reservoirs can provide the necessary storage but it is unlikely to be cost effective to construct on-farm storage areas on the majority of the scattered farms in Botswana. For boreholes, treated wastewater, river and sand river withdrawals, the highest daily yield will constrain the irrigable area. Monthly water requirements for various NAMPAADD scenarios are given in Figure 8.1. Generally the scenarios showed similar water demand.

The total planned irrigated area under horticultural and field crops in NAMPAADD under the economic and social scenario is 5,200 and 5,400 ha, respectively. Of this 1,600 ha in the economic scenario and 1,900 ha in the social scenario was planned for irrigation with mainly groundwater, complemented by some river abstractions. The remaining area will be irrigated by treated wastewater. Table 8.3 presents the summary of potential production and planned utilisation for irrigation for the two NAMPAADD scenarios.

Table 8.3: The potential production and planned irrigation for NAMPAADD scenarios in Ml/year

	Existing boreholes	Existing dams	Existing stream-flow	Potential ground-water	Potential waste-water	Total
Potential production	19622	1694	5102	22840	47500	96758
Planned utilization by 2010 (socially oriented)	8210	2482	3496	1453	35047	50688
Planned utilization by 2010 (Econ. Oriented)	6134	2058	3445	1042	35047	47726

Source: NAMPAADD quoted in SMEC 2006

Sources of water for irrigation

Groundwater (including sand river water abstractions), river withdrawals, wastewater and small dams comprise the main source of water for irrigation. Monitoring of these sources is not conducted systematically and therefore there is a lack of reliable data on quantity and quality of water used.

Groundwater

Groundwater constitutes the main source of water for irrigation. The registry at Department of Water Affairs (DWA) records a total of over 11,000 permits issued for groundwater abstraction, which corresponds to an estimated 11,500 Ml/year. The actual quantity extracted may be higher as there are no controls or policing of abstractions.

Surface water

Surface water supplies for irrigation are limited to abstraction from flowing rivers and streams, small and medium livestock dams, and recycled wastewater. Availability of river water flows is limited as Botswana has only three perennial rivers, the Limpopo, the Okavango and the Chobe. These rivers are subject to international protocols that have not been finalised. Other rivers are ephemeral, with the majority having no flows at all.

Although the permits to pump water specify the maximum amount of water that can be extracted, these limits are not adhered to due to lack of monitoring and accurate data on surface water usage (SMEC 2006). NAMPAADD projected that by 2010, the potential availability of water for irrigation in rivers would be 7,118Ml/year. The Water Development Section (WDS) of the Ministry of Agriculture (MoA) builds, operates and maintains approximately 800 small and medium livestock dams throughout the country. The primary function of these dams is to provide storage for livestock watering but surplus water may be available for ir-

Table 8.4: Small dams currently used for irrigation

Dam	Location	Approximate area that can be irrigated (ha)
Mmamokhasi	Kanye	20
Kubung	Kubung	8
Somanka	Malotswane	5
Dikabeya	Palapye	60
Mmakgodumo	Kanye	8
Semarule	Molepolole	18
TOTAL		119

Source: Data from Ministry of Agriculture cited in SMEC 2006

rigation. The WDS initially identified about 15 dams with potential for irrigation but presently there are only seven dams with potential for irrigation (Table 8.4).

Reclaimed wastewater

Reclaimed wastewater offers an excellent source of water for irrigation in Botswana. The Botswana National Master Plan for Waste Water and Sanitation (NMPWWS) of 2003 states that only 10% of the return flow from wastewater treatment plants is reused, the rest being lost to seepage and evaporation. The NMPWWS of 2003 further estimated that 12,000 Ml/year of treated wastewater is available in Botswana; 80% of this is from the five major urban centres of Gaborone, Francistown, Selibe-Phikwe, Lobatse and Orapa.

Current irrigation in the Okavango Delta

Water for various human uses is widely available from the vast floodplains of the Okavango Delta. However, water of the quality required for human consumption and other uses is supplied by the Water Utilities Corporation. This parastatal organisation is currently overwhelmed with supply challenges associated with burgeoning demand. Some studies have revealed potable water supply challenges, including water shortages, poor water quality and difficulties in the collection of water fees (Mazvimavi and Mmopelwa 2006; Kgomotso and Swatuk 2006). In addition to national legislation and the water policy framework embodied in the National Water Master Plan Review of 2006, water resources management in the Okavango Delta is supported by the ODMP of 2008. This is particularly so regarding the improvements in water management institutions to enhance water quality monitoring (Department of Environmental Affairs 2008). However, the plan does not provide guidelines for irrigation management, and only briefly refers to land use zones as stated in the Ngamiland Settlement Strategy of 2003, particularly

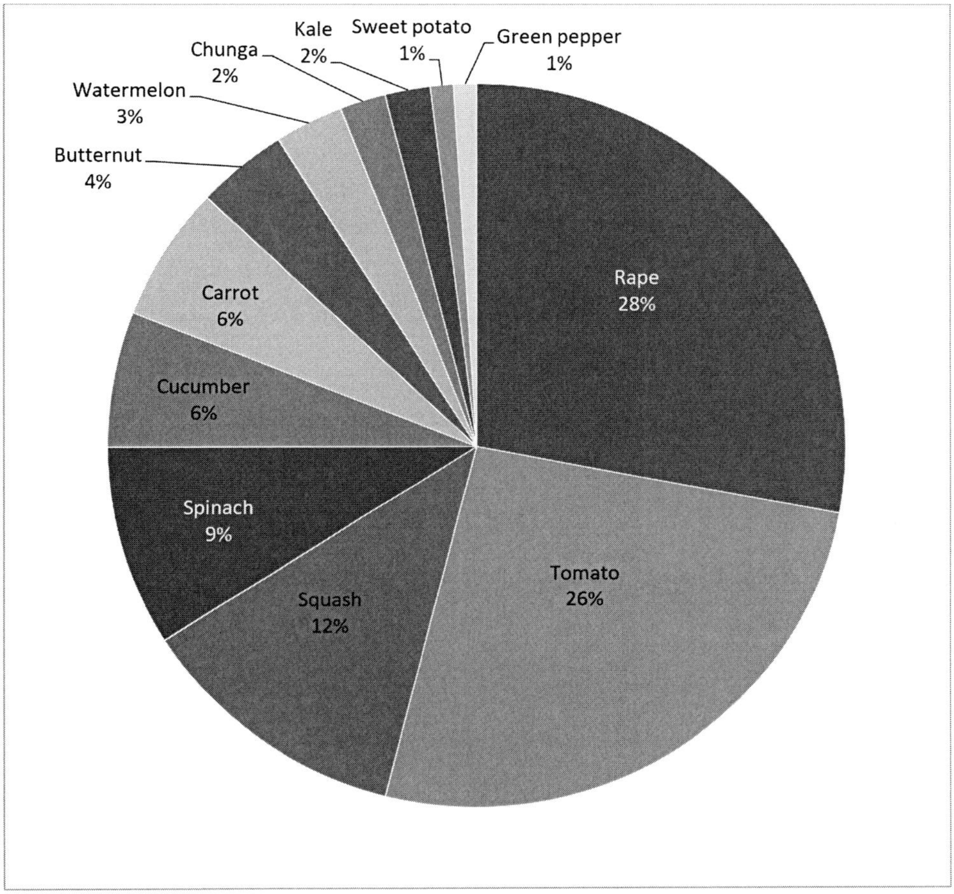

Figure 8.2 Proportion of area used to grow each vegetable type in Ngamiland East.

the potential for irrigation in the Maun Zone (which includes the villages of Sehitwa, Matlapana, Shorobe and Makalamabedi). Irrigation is also subsumed, in the Ngamiland District Development Plan, as part of the nationally driven NAMPAADD programme. ODMP implementation is not the task of a specific department; it is dependent on the implementation of other plans (such as the NDP 10 and the Ngamiland District Development Plan 7), and it does not have an autonomous time frame. Therefore, it lacks direct focus on irrigation with particular reference to the Okavango Delta.

This major omission may have been due to the fact that the ODMP planning process was issue-driven rather than broad based and forward looking. Irrigation was therefore not indicated as one of the major concerns during the planning period. Currently, and this was the case during the planning period, there is little irrigation being undertaken around the Okavango Delta and Ngamiland District.

From the data obtained from the Regional Agriculture Office and Department of Agricultural Research in Maun, of 188.15 hectares of land allocated for irrigation, only 31.35 hectares (17%) is being utilised. Most of the irrigation is for vegetables. All the irrigation farms are owned by citizens except for one commercial farm which covers 81.25 hectares, of which only 1.25 hectares is currently being utilised. The general distribution of land demarcated for irrigation in various areas of Ngamiland district is indicated in the following sub-sections.

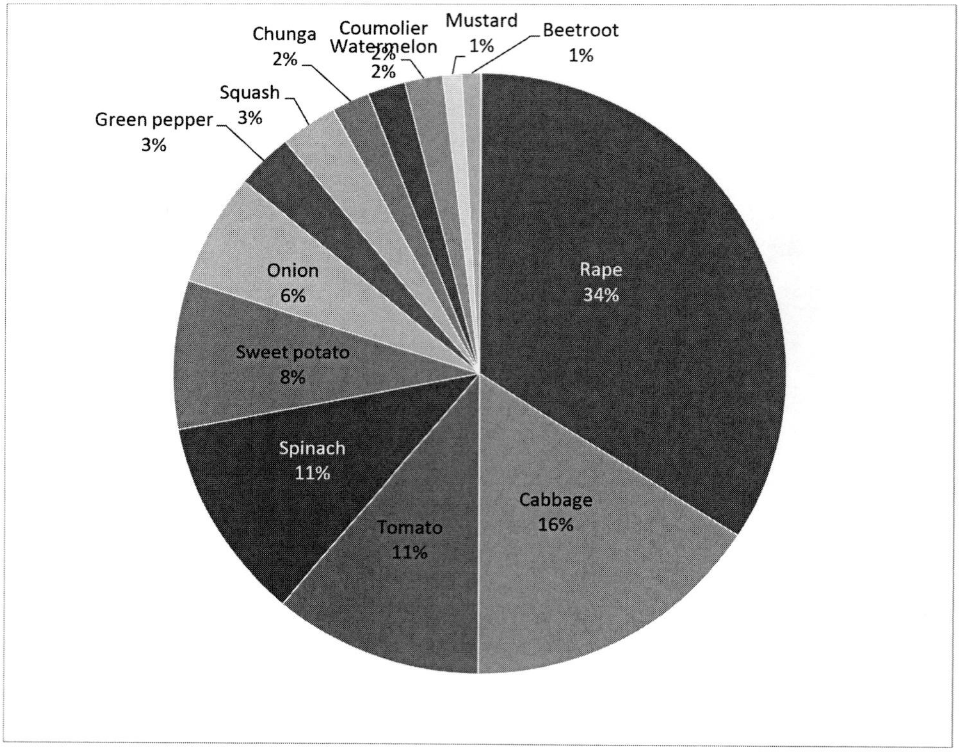

Figure 8.3: Proportion of area under each vegetable in the Okavango.

Ngamiland East (Maun)

The allocated land for irrigation is 48.8 hectares. Only 15.2 hectares of these (31%) are under irrigation (Table 4). The mean and mode for the irrigated land is 0.46 hectares and 0 hectares, indicating that many of the allocated plots are not being irrigated. All the irrigation is for vegetables except for two farms. The proportion of the area under each vegetable grown is given in Figure 8.2. Rape and tomatoes are the most prolific crops, covering 28% and 26% of the area under vegetables respectively. Squash (12%), cucumber (9%) and spinach (6%) are next

Table 8.5: Irrigation projects in Ngamiland East District (Maun).

LOCATION	SOURCE OF WATER	IRRIGATION SYSTEM	CROPS GROWN	LAND ALLOCATED (HECTARES[1])	IRRIGATED AREA (HECTARES)
Bodibeng	River	Stand pipe	Vegetables	1 (1)	0
Boro	Borehole	Microjets		1 (1)	0
Kareng	Borehole	Stand pipe		1 (1)	0
Makgalong	Open well, borehole	Sprinkler	Vegetables	1 (1)	1
Matlapana	River, river well point	Sprinkler	Vegetables	4.5 (3)	0.03
Maun	Borehole, River	Sprinkler, tunnels, drip	Vegetables	6.8 (3)	1.45
Mawana	Borehole	Sprinkler		1 (1)	0
Mphampha	Borehole	Microjets	Citrus	2.5 (2)	2.5
Nxaraga	Borehole/ river	Stand pipe, drag hose, sprinkler, microjets	Vegetables/ fruits	4 (4)	3
Roman	Borehole	Hose		1 (1)	–
Samedupi	River	Sprinkler/ microjets	Vegetables	6.5 (5)	3.42
Semboyo	Borehole	Sprinkler	Vegetables	1 (1)	0
Sexaxa	River	Sprinkler	Vegetables	1 (1)	0.5
Shorobe	Borehole	Stand pipe	Vegetables	1 (1)	1
Tsanekona	River	Sprinkler	Vegetables	1 (1)	0.5
Tsao	Borehole	Sprinkler	Vegetables	1 (1)	0
Xhobe	River	Sprinkler/ microjets	Vegetables	13.5 (7)	1.8
TOTAL				48.8 (35)	15.2

Source: Ministry of Agriculture, Crop Production Section, Maun. The number of projects in the village is indicated in parenthesis.
[1]The number of projects is indicated in parenthesis.

in importance in terms of the proportion of the area under cultivation. The rest as demonstrated in Figure 8.2 are grown on a much smaller area.

Ngamiland West (Gumare)

Only 20 hectares are allocated for irrigation in Ngamiland West, and of these, 8.5 hectares (43%) are being utilised (Table 8.5). The mean and maximum area utilised per farmer are 0.6 hectares and 1.5 hectares respectively. Six of the 15 farms in this area use open wells and the rest use river water. The breakdown of area under different vegetables was not available.

Okavango

A total of 119.35 hectares have been allocated for irrigation in the Okavango. Out of these, one is a commercial farm that was allocated 81.25 hectares. Only 6%

of the allocated land (7.65 hectares) is being utilised (Table 8.6). Only one farm uses a borehole whereas others use river water. Figure 8.3 shows the proportion of the area cultivated under each vegetable. Rape (34%), cabbage (16%), spinach (11%) and tomato (11%) are the vegetables grown the most. Unlike Ngamiland East, cabbages are more significant in the Okavango area.

Table 8.6: Irrigation projects in Ngamiland West (Gumare).

LOCATION	SOURCE OF WATER	IRRIGATION SYSTEM	CROPS GROWN	LAND ALLOCATED (HECTARES[1])	IRRIGATED AREA (HECTARES)
Tubu	Open well	Stand pipe/ bucket	Vegetables	4 (5)	2
	River	Hose	Vegetables	1 (1)	0.75
Sepopa	River	Stand pipe/ Sprinkler	Vegetables	3.5 (3)	2
Etsha 6	River	Sprinkler/hose/ bucket	Vegetables	8 (4)	1.75
Etsha 13	River	Bucket	Vegetables	1 (1)	0.5
Danega	Open well	Stand pipe	Vegetables	1 (1)	0
Konde	Open well	Hose	Vegetables	1.5 (1)	1.5
TOTAL				20 (16)	8.5

Source: Ministry of Agriculture, Crop Production section, Maun.
[1]The number of projects is indicated in parenthesis.

Table 8.7: Irrigation projects in Okavango area

LOCATION	SOURCE OF WATER	IRRIGATION SYSTEM	CROPS GROWN	LAND ALLOCATED (HECTARES[1])	IRRIGATED AREA (HECTARES)
Gunitsoga	River	Other	Vegetables	3.5 (1)	-
Seronga	River	Sprinkler, other	Vegetables	4.4 (2)	1
Kauxwi	Stand pipe	Hose	Vegetables	0.2 (1)	0.2
Mohembo	River	Hose/other	Vegetables	7 (2)	0
Shakawe	River	Sprinkler, basin, centre pivot, hose	Vegetables	97.5 (10)	5.7
6. Samochima	River	Sprinkler, mi-crojets	Vegetables, citrus	5 (2)	0.5
7. Nxam-asere	River	Sprinkler	Vegetables	0.75 (1)	0.25
8. Beetsha	Borehole	Hose	Vegetables	1 (1)	0
TOTAL				119.35 (20)	7.65

Source: Ministry of Agriculture, Crop Production section, Maun.
[1]The number of projects is indicated in parenthesis.

Water Usage

Current and short-term future water usage

Based on the allocated and actual area under irrigation, the current and future short-term water usage was estimated using the pan evaporation values of Maun and Shakawe/Gumare (Table 8.7). Irrigation is mostly carried out during the dry months of April to October hence the current annual water usage in Ngamiland East is about 264,000 m^3. If all the land currently allocated were used, the future short-term annual water usage would be 837,000 m^3.

A similar estimate for Ngamiland West and Okavango combined shows that the current annual water used in irrigation is 256,000 m^3 and the short-term future requirements will be 2,206,000 m^3 (Table 8.8). The current and short-term use of the water is therefore small.

Table 8.8: Current and future water for irrigation in Ngamiland East

MONTH	EVAPORATION	ESTIMATED AMOUNT OF WATER NEEDED FOR IRRIGATION		
		m^3/ha/month	Total[1] (m^3) for 15.2 Ha	Total[2] (m^3) for 48.3 Ha
April	199	2274	35,000	110,000
May	189	2160	33,000	104,000
June	158	1806	27,000	87,000
July	183	2091	32,000	101,000
August	218	2491	38,000	120,000
September	263	3006	46,000	145,000
October	307	3509	53,000	169,000
Total (m^3/yr)			264,000	837,000

[1]Current water usage.
[2]Future short term estimated water usage in irrigation.

Long-term water usage

To date, Botswana has not planned any major irrigation activities along the Delta, while international protocols are being finalised. However, based on the land suitability map (see Figure 8.4), most of the area around the Okavango Delta could result in moderate to moderately high yields in rain-fed agriculture. The yields would improve if irrigation was encouraged and supported.

Issues of trans-boundary importance

Since Botswana is downstream of the Okavango River Basin, there is a need to consider the issues of trans-boundary significance with regard to irrigation. Currently, there is limited irrigation in the upstream riparian countries of Angola and

Table 8.9: Current and future water for irrigation in Ngamiland West and Okavango

MONTH	EVAPORATION	ESTIMATED AMOUNT OF WATER NEEDED FOR IRRIGATION m³/ha/month	Total[1] 16.15	Total[2] 139.35
April	192	2194	35,000	306,000
May	178	2034	33,000	283,000
June	146	1669	27,000	233,000
July	161	1840	30,000	256,000
August	190	2171	35,000	302,000
September	241	2754	44,000	384,000
October	277	3166	51,000	441,000
Total (m^3/yr)			256,000	2,206,000

[1]Current water usage.
[2]Future short term estimated water usage in irrigation.

Namibia. However, there may be a need to use the water resources of the Okavango River and the related catchment to supply water for various needs including large-scale irrigation in all the riparian countries. Water extraction for irrigation and pollution from the use of pesticides and fertilisers are possible threats to the availability of water for environmental and other requirements.

Water abstraction

Due to climatic variability in the southern African region, water availability in the Okavango River Basin is also variable. As a result there are uncertainties resulting in general fear that extensive abstraction of water upstream may result in low flows into the Okavango Delta in Botswana. The availability of water for large-scale irrigation may also be subject to this variability, especially considering that irrigation would be competing with other uses. This is also true for water abstraction in the northern part of the Delta affecting the downstream areas. The environmental flow requirement of 390m^3/s stipulated in the National Water Master Plan Review (Volume 6) needs to be evaluated to check whether it could be adapted for determining water abstraction upstream.

Pollution

Whereas water for irrigation may not need to be of high quality, certain quality parameters may affect whether water can be used for irrigation, including the concentration of pollutants.

Pollution resulting in high salinity

A salinity problem occurs if salts accumulation in the crop zone is concentrated enough to cause a loss in yield. These salts may originate from salts in the applied

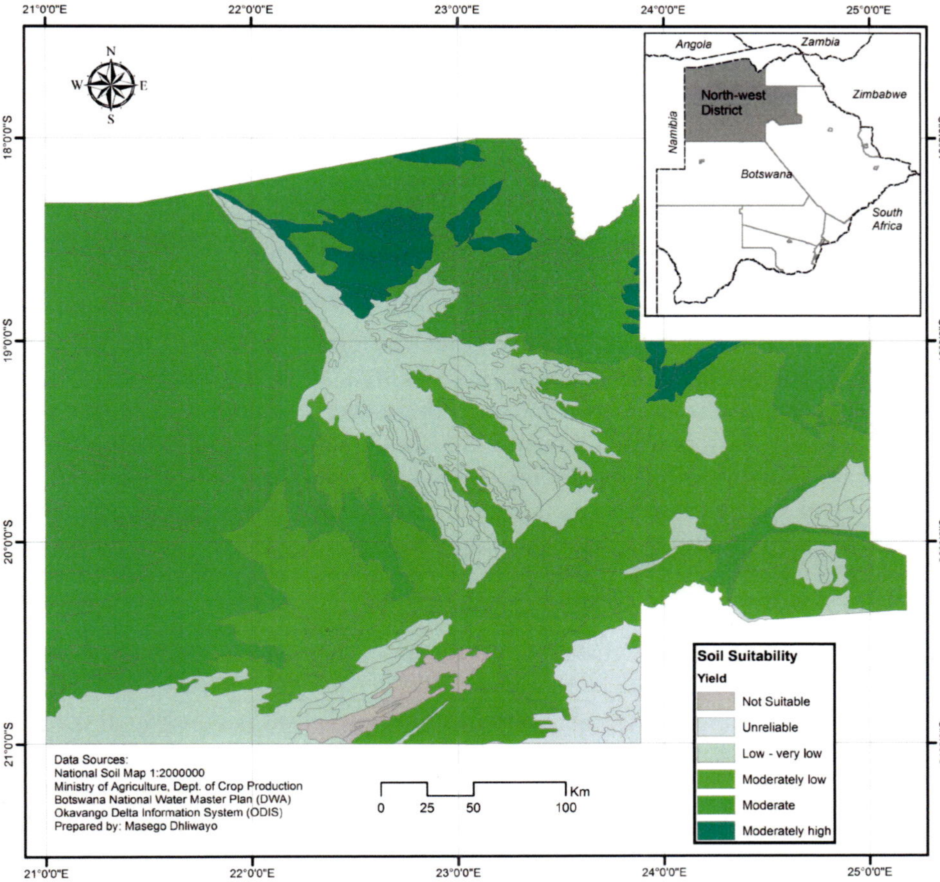

Figure 8.4: Land suitability map of Ngamiland. Source: ORI GIS Laboratory

Table 8.10: Pesticides and fungicides recommended for vegetables by the Department of Agricultural Research

Pesticide/fungicide	Pesticide/fungicide	Pesticide/fungicide
Mancozeb	Permethrin	Cypermethrin
Copper oxychloride	Apron C	Deltamethrin
Funginex	Ridomil	Gamma-BHC
Ridomil	Bravo 50	Nemacur
Apron C	Dithane M45	Thiram
Dichlorvos		
Carbofuron		

water. Currently the main-channel water of the Okavango is of low salt content; but high salt levels are observed in some pools. Any process that increases the salt content of the water has the potential to contribute to the salinity problem. Such processes include discharge of wastewater from a reverse osmosis plant, or discharge of sewage effluent into the water.

Pollution resulting in low water infiltration rates

An infiltration problem related to water quality occurs when the normal infiltration rate for the applied water or rainfall is reduced appreciably and water remains on the soil surface for too long or infiltrates too slowly to supply sufficient water to maintain acceptable crop yields. Although the infiltration rate of water into soil varies widely and can be greatly influenced by the quality of the irrigation water, soil factors such as structure, degree of compaction, organic matter content and chemical make-up can also influence the intake rate greatly. The two most common quality factors which influence the normal infiltration rate are the salinity of the water and its sodium content relative to the calcium and magnesium content. High salinity water will increase infiltration. Low salinity water or water with high sodium to calcium ratio can decrease infiltration.

Pollution resulting in toxicity

Toxicity problems occur if certain constituents in the water are absorbed by the plant and accumulate to concentration levels high enough to cause crop damage – of primary concern are chloride, sodium and boron. Introduction of high concentrations of these into the water upstream of Botswana will impact negatively on irrigation.

Based on information from the Department of Crop Production, there is very little use of fertilisers and pesticides in the areas around the Okavango Delta. In some cases, livestock manure is used but the amounts used are not documented. Certain pesticides and fungicides (Table 8.10) are recommended by the Department of Agricultural Research for use in vegetable production farms in Botswana and Ngamiland District in particular. However, monitoring of the use of such chemicals is still rudimentary, and could pose a potential risk if the chemicals are used in quantities which may not be environmentally sustainable. Currently the use of these chemicals in Ngamiland is regarded as very low given that there is limited commercial crop production.

Conclusion

This chapter has shown that agricultural activities contribute 2% to Botswana's GDP, 80% of which derives from the beef sector. Botswana is a net importer of agricultural products including cereals and vegetables. The country faces major challenges to arable production, including generally infertile soils and low rainfall. However, even in potentially productive regions such as the Chobe and Ngamiland Districts, arable production does not fully optimise the comparative advantage in the form of large surface water resources. Only 1,800 hectares are currently under irrigation in Botswana, with plans to increase this to 5,400 hectares in the short term and up to 40,000 hectares in the long term. Long-term irrigation is planned for the Pandamatenga-Chobe area with water drawn from the Chobe/Zambezi River system. However, the implementation of this plan is subject to international agreements.

The chapter has further revealed that 31.38 hectares out of 118.15 hectares of land near the Okavango Delta are under irrigation. There are no plans for extensive irrigation using waters from the Okavango Delta until the conclusion of international agreements on the use of the water from the Okavango. From the foregoing it can be deduced that in the Ngamiland District, arable production through irrigation faces institutional constraints which may effectively reduce the potential to optimise the District's comparative advantage over other regions in Botswana. This is a major challenge for food security in the country given that most river catchments in Botswana are in internationally-shared watercourses. Coupled with other factors such as population growth and climate change, institutional constraints are likely to result in adverse impacts on poverty reduction, particularly among the rural poor.

For planning purposes, an environmental flow requirement with an accompanying definition of the acceptable ecological limits for the Okavango Delta should be undertaken as a matter of urgency. The environmental flow requirement is currently being undertaken but there is need to define acceptable ecological integrity that balances environmental and developmental needs. This would provide a measure of how much water can be drawn from the Okavango system for irrigation. There is need to have irrigation plans within the broader IWRM strategy for the Okavango Delta. The plans should pay attention to and provide tools to monitor the use of fertilisers, manure and pesticides in the Okavango Delta in order to minimise their adverse effects.

References

Department of Crop Production. (2006). Draft Botswana Irrigation Policy and Strategy, unpublished.

Department of Environmental Affairs. (2008). Okavango Delta Management Plan, Gaborone: Department of Environmental Affairs.

Central Statistics Office. (2012). *National Accounts Statistics Report 2011*, Central Statistics Office, Gaborone.

Central Statistics Office. (2008). *National Accounts Statistics – Quarterly Gross Domestic Product*, Fourth Quarter 2008.

Ministry of Finance and Development Planning (2009). *National Development Plan 10,* Gaborone: Ministry of Finance and Development Planning.

Ministry of Agriculture (n.d.). National Master Plan for the Arable Agriculture and Dairy Development [NAMPAADD] guidelines. Gaborone: Government Printer.

Ministry of Finance and Development Planning (2003). *National Development Plan 9 2003/04–2008/09*, Gaborone: Ministry of Finance and Development Planning.

Pike, J.G. (1970). "Interim Technical notes on the hydrology and water resources of the Okavango River system", FAO and Government of the Republic of Botswana.

Randall, D. (1957). "Factors of Economic Development and the Okavango Delta", University of Chicago Programme of Education and Research Planning Research Paper No. 3; Department of Geography Research Paper No. 47.

Rahm, R., Swatuk, L. and Matheny, E. (2006). "Water resource management in Botswana: balancing sustainability and economic development." *Environment, Development and Sustainability*, Vol. 8, 157–183.

Scudder, T., Manley, R.E., Coley, R.W., Davis, R.K., Green, J., Howard, G.W., Lawry, S.W., Martz, D., Rogers, P.P., Taylor, A.R.D., Turner, S.D., White, G.F., and Wright, E.P. (1993). The IUCN review of the Southern Okavango Integrated Water Development Project, IUCN.

Snowy Mountain Engineering Corporation Ltd [SMEC] (2006a). National Water Master Plan Review Final Report Volume 1. Main Report, Ministry of Minerals, Energy and Water Resources, Gaborone.

SMEC (2006b). National Water Master Plan Review Final Report, Volume 8: Agriculture. Ministry of Minerals, Energy and Water Resources, Gaborone.

SMEC, WLPU Consultants and Swedish Geological International AB. (1991). *Botswana national watermaster plan study.* Final Report, Volume 8, Rural Water Supply, Gaborone, Botswana: Ministry of Mineral Resources and Water Affairs, Department of Water Affairs.

SMEC (1987). Southern Okavango Integrated Water Development Project, phase I: final report technical study [5 volumes], Report to DWA by SMEC, June 1987.

Southern African Development Community [SADC] (2006). Regional Water Strategy, Gaborone, Botswana: SADC.

Statistics Botswana (2011). *Botswana core welfare indicators(Poverty) survey of 2009/10 preliminary results*, Gaborone: Statistics Botswana.

Statistics Botswana (2012). *National Accounts Statistics – gross domestic product: fourth quarter and annual 2011*, Statistics Brief No 2012/4, Gaborone: Statistics Botswana.

TAHAL Consulting Engineers (2000). National Master Plan for the Arable Agriculture and Dairy Development Final Report, Volume 4, Climate, Soils and Water.

The Permanent Okavango River Commission [OKACOM] (2011). Strategic action programme (SAP) for the sustainable development and management of the Cubango-Okavango basin, OKACOM Maun, Botswana.

The World FactBook. (n.d.) https://www.cia.gov/library/publications/the-world-factbook/geos/bc.html#Econ. Accessed 21 May 2009.

Swatuk, L.A., and Kgomotso, P.K. (2007). "The challenges of supplying water to small, scattered communities in the Lower Okavango Basin, Ngamiland, Botswana: an evaluation of government policy and performance", *Physics and Chemistry of the Earth*, 32, (1264–1274).

Swatuk, L.A. and Rahm, D. (2004). "Integrating policy, disintegrating practice: water resources management in Botswana", *Physics and Chemistry of the Earth*, 29, 1357–1364.

Van der Sluis, T. (1990). "Irrigation zoning Nhabe River Ngamiland", Regional Agriculture Office/DLUPU, Maun.

9 Small-Scale Fisheries Management in Floodplain Systems: The Case of the Okavango Delta. What Needs to be Managed? How Should it be Managed?

Ketlhatlogile MOSEPELE

Abstract

Fisheries management is an interdisciplinary subject and an integral part of natural resources management dealing with fisheries as a complex system of interaction between humans and nature. Berkes (2003) argues that the philosophical basis of resource management, which forms the basis for fisheries management, is treating people and nature as commodities in the service of efficient and often parasitic exploitation for industrialisation, capitalism and colonialism. Possibly due to the high dependence of fishers on small-scale fisheries (SSF), most studies have used Hardin's (1968) "tragedy of the commons" scenario to argue that the open access nature of these fisheries has resulted in resource degradation, poverty and marginalisation (Allison and Ellis 2001); hence the need for management. The aim of this study is to determine the impact of classical management paradigms on floodplain fisheries of the Okavango Delta. The study relies on a literature synthesis of conceptual issues and an analysis of the perceptions of small-scale commercial fishers of the Okavango Delta Panhandle. It found that current fisheries management in Botswana is indeed based on the classical model which is however not suitable for dynamic floodplain fisheries. It was further found that the adopted management system is both a potential ecological and livelihood problem for the SSF of the Okavango Delta. The chapter concludes by recommending a holistic and adaptive fisheries management system where the ecosystem needs are integrated into social needs of the fishing community and are based on the framework of sustainable development.

Keywords: adaptation, environmental variability, indigenous, management, SSF

Introduction

There is no universally accepted definition of small-scale fisheries, because these vary across different spatial and temporal scales (Staples *et al.* 2004; FAO 2005; Carvalho *et al.* 2011), though Kolding *et al.* (2013) provide a succinct definition that encapsulates the character of SSF at a global scale. According to Carvalho *et al.* (2011), SSF use low technology fishing gear compared to large-scale fisheries. SSFs are diverse in nature, and are highly adaptable to different environments, resources, and seasons, which makes them more attuned to the character of the local ecosystem (Kolding *et al.* 2013). However, for the purposes of this study, SSF are defined as dynamic entities characterised by labour-intensive activities (harvesting, processing and distribution), targeting multi-species, using various fishing gear and techniques, most of which are relatively simple in a multi-use, multi-user environment (FAO 2005). Therefore, SSF include traditional, artisanal and subsistence fisheries (Berkes 2003).

Many small-scale fishing communities are poor and vulnerable (FAO 2005), and are characterised by remoteness, poor infrastructure, weak political power and seasonality (Campbell and Townsley 1996; Jacquet and Pauly 2008). Small-scale fisheries contribute more to local economies (FAO 2005) while large-scale fisheries contribute more to the GDP of national economies (Hall and Andrew 2011). The lack of economic power of SSF means they command very little of the political power essential for their development as sources of sustainable livelihoods (Staples *et al* 2004). Of 51 million fishers globally, only 500,000 of them do not work in SSFs (Berkes 2003), making SSFs a major source of employment. There is a disproportionately higher proportion of women involved in SSFs than their male counterparts (Hall and Andrew 2011), suggesting that a comprehensive management of this resource could be an important pathway towards the socio-economic empowerment of women. This observation is supported by Welcomme *et al.* (2010) who observed that 54% of the people involved in the post-harvest sector of inland SSF are women.

Potentially, SSF are more sustainable than large-scale fisheries (Jacquet and Pauly 2008) because of their low technology usage (Kolding *et al.* 2013). Furthermore, globally SSF are the sixth major source of animal protein, and the fourth major source in Africa (Welcomme 2011), which makes them a key source of livelihoods for socio-economically marginalised riparian communities. Therefore, SSF contribute significantly to food and nutrition security, to sustainable livelihoods, and to poverty alleviation in most socio-economically marginalised riparian communities at a global scale (Staples *et al.* 2004). Despite these positive attributes, a dearth of knowledge and information about SSF has resulted in them being marginalised or ignored often in national and regional planning and policy development (Staples *et al.* 2004). Often SSF are placed under inappropriate management emanating from top-down conservation ideas and ideals.

What is fisheries management and why is it necessary?

Fisheries management is an interdisciplinary subject dealing with fisheries as a complex system of interaction between humans and nature. Berkes (2003) argues that the philosophical basis of resource management, forming the basis for fisheries management, is treating people and nature as commodities in the service of efficient and often parasitic exploitation for industrialisation, capitalism and colonialism. Possibly due to the high dependence of fishers on SSF, most studies have used Hardin's (1968) "tragedy of the commons" scenario to argue that the open access nature of these fisheries has resulted in resource degradation, poverty and marginalisation (Allison and Ellis 2001). This high dependence in SSF is argued to have resulted in either "perceived" (Welcomme *et al.* 2010) or observed (Allan *et al.* 2005) declines in fish catches from inland fisheries due to an open access regime which has caused unsustainable exploitation levels (Viner *et al.* 2006).

In this scenario, management aims are twofold: to rebuild over-exploited fisheries (Branch *et al.* 2012) and to ensure sustainable utilisation. Andrew and Evans (2009) highlight what they term "perspectives" on fisheries management: (1) *"inside looking in"* which is the classical view of fisheries management where threats and solutions come from within the fishery; (2) *"inside looking out"* which essentially suggests that threats and opportunities come from outside the domain of fisheries; (3) *"outside looking in"* which mainstreams fisheries management and governance; and (4) *"outside looking out"* which suggests that fisheries per se are not important in themselves due to a failure in investment in SSF. These perspectives encapsulate what Andrew and Evans (2009) consider to be the three key approaches to fisheries management; (i) ecosystem based management; (ii) rights based management; and (iii) management for resilience.

The classical perspective of fisheries management is that fisheries are managed for the benefit of mankind (Hillborn and Waters 1992). Profit optimisation and cost reduction are the key imperatives of fisheries management, where the key management objective of fisheries is to harvest fish stocks at the maximum sustainable yield (MSY) (EU 2006). According to Caddy (1996), declining fish yields and over-capitalisation of global marine fisheries indicate a clear and present need for fisheries management. Alan *et al.* (2005) also argue that SSF are being over-exploited, which necessitates the need for urgent management intervention. To sum up the main argument of the classical view, open access regime causes intense competition (Craig *et al.* 2004; Allan *et al.* 2005) which invariably results in excessive fishing (Perez *et al.* 2012), resource degradation, poverty and marginalisation (Allison and Elis 2001).

The FAO 2005 report presents a far clearer reason why management is important. It submits that fish are not an infinite resource, hence the need for their management to ensure sustained production. Yet another argument is that essentially the burgeoning global demand for fish has resulted in crises and controversy (Delgado *et al.* 2003) which, despite management regulations based on classical approaches, has resulted in management failure (Baelde 2003). This management failure is based on two fundamental concerns: (a) that classical fisheries management de-links ecology from social issues when the two are intricately linked; and that (b) classical approaches manage for products and commodities instead of managing for resilience (Berkes 2003). The understanding, of course is that, if managed properly SSF can generate significant profits, provide resilience to shocks and crises, and can contribute to poverty alleviation and food (FAO 2005) and nutrition security (Nnyepi *et al.* 2007) among riparian communities. Unfortunately, fisheries management interventions are implemented on the background of perceptions or fears of over-fishing, especially in the so-called open access fisheries (Kolding *et al.* 2013).

Classical fisheries management (how and who?)

Classical fisheries management is a mixture of biology and economics, developed from single species fisheries, with a western and northern flair (Berkes 2003). According to Welcomme (2001), classical fisheries management in floodplain fisheries comprises technical measures (e.g., mesh restrictions, gear restrictions, closed seasons, closed areas), input controls (e.g., licenses, state regulated access, ownership), and output controls (e.g., quotas, size limits). The classical management paradigm includes not only technical controls of the fishery, but also power structures in the management model (Nielsen *et al.* 2004).

Nielsen *et al.* (2004) argue that modern fisheries management is based on centralised government approaches which make it impossible to deal with contemporary management concerns. These include, among others, the effects of globalisation which requires the "integration of local markets into global markets and the subsumption of political and social processes under international economic forces, which often leads to exclusion rather than new opportunities for fishing communities" (Nielsen *et al.* 2004: 151). Nielsen *et al.* (2004) further highlight that centralised government management of fisheries makes it difficult to deal with other issues such as resource user conflict and prevention of resulting social disruption. In essence, classical management approaches invest power in centralised government entities, where local/ riparian communities are essentially disempowered. This management model is based on command and control structures, where authority and knowledge flow from a top-down approach.

Nielsen *et al.* (2004) highlight that the challenges facing fisheries are multidimensional and multi-scale; some of the key governance issues relate to the globalisation of fisheries issues through international agreements and conventions that focus on the ecosystem as a whole and not specifically on fisheries. These invariably affect fisher access to the fishery resource. Other uses of water resources such as infrastructural development like dams, irrigation schemes, among others, also place the fishery resources under tremendous pressure. Invariably, these affect the integrity of these resources which ultimately affects fisher access to the resource base.

Alternative management regimes (how and who?)

Hoggarth *et al.* (2006) highlight that SSF management needs more involvement by key stakeholders, transparency, decentralisation of decision-making, and coordination of inter-sectoral linkages between fisheries and the wider social and ecological systems. The broad spectrum of fisheries management paradigms ranges from loose alliances to more rigid organisational structures (Perez *et al.* 2012). Allison and Ellis (2001) propose the livelihoods approach as an alternative approach to management of SSF. Allison and Ellis (2001) argue that this approach incorporates uncertainty and is characteristic of SSF, hence making it a more appropriate management approach of these fisheries. According to Kolding *et al.* (2013), self-government of SSF is the most appropriate management paradigm. They argue that self-government activities enjoy greater legitimacy than dictated policies (from top-down management approaches) and stand a better chance of success, especially in weak political systems. This argument gains traction from research showing that communities have the ability for self-organisation and self-regulation to solve the exclusion and extractability of the commons (Berkes 2006). On the other hand, Staples *et al.* (2004) argue that alternative approaches to management of SSF should include elements such as the ecosystem approach, integrated resource management, decentralisation and community participation, user and access rights, and market-based approaches.

Co-management is seen as the foundation for alternative fisheries management regimes such as integrated area management, ecosystem based management, protected area management and adaptive management (Evans *et al.* 2011). It is perhaps based on this philosophy that some studies (Hoggarth *et al.* 1999) argue for an integrated fisheries management approach in SSF due to the plethora of issues facing inland fisheries. These regimes (i.e. co-management) in SSF have been used extensively around the world with much success, for example in Bangladesh (Craig *et al.* 2004), Cambodia (Viner *et al.* 2006; Fennel *et al.* 2008), Botswana (Mosepele *et al.* 2014) and the Amazon (Pinho *et al.* 2012). In fact, Evans *et al.* (2011) highlight that co-management regimes are the dominant approach to management. However, their efficacy in the management of SSF is questionable (Pomeroy *et al.* 2001; Chuengpagdee and Jentoft 2007) given the issue of scale. Berkes (2006) argues that the fundamental challenge with management of the commons (SSF) is that success stories of co-management are based on local scale studies, and that external and internal challenges might cause these to fail when up-scaled to the bigger picture of SSF mismanagement. Moreover, Chuengpagdee and Jentoft (2007) discuss that the success of co-management of SSF depends on the pre-implementation of the process. Their argument is that co-management that goes wrong in the initial steps would most likely fail. On the other hand, Gutierrez *et al.* (2010) contend that leadership is a key variable that determines the success and/or failure of co-management regimes. Gutierrez *et al.* highlight that a highly motivated community leader with strong entrepreneurial skills, who is committed to the co-management process, is sometimes the bridge between success and failure of these regimes.

The issue of holistic management, where fishers are also part of the governance structure of fisheries, entails integrating both the fisher knowledge and including fishers in the management regime. Fishers are a valuable source of knowledge on fisheries management because they are, in fact, the only available source of local, historical, place-based fisheries information (Ames 2003). Several studies from around the world highlight indigenous knowledge as a key tool in attaining holistic fisheries management. Native American fishers use traditional indigenous knowledge to target their preferred species (Narcisse 2003), and to develop adaptive fishing gear to optimise their catches due to variations in fish species availability (Lucas 2003). Essentially, management of SSF is a delicate balancing act that involves six core elements: sustain, protect, develop, capacitate, institutionalise and communicate (Perez *et al* 2012).

Christie *et al.* (2007) highlight ecosystem based fishery management (EBFM), which postulates that sustainable fisheries can be achieved through a management regime that integrates fish stock interactions, predator-prey interactions and fish species habitat requirements. Christie *et al.* (2007) further argue that EBFM can be achieved if it incorporates the regulation of fishing effort and its effects on fish stocks. Adaptive co-management has also been seen as a panacea to SSF management. It starts from a perspective of incomplete information, and essentially develops a management approach based on trial and error, before a final management decision is made (Berkes 2003). The aim of this study as presented in this chapter is to determine the impact of classical management paradigms on floodplain fisheries of the Okavango Delta.

Materials and methods

Study area

The Okavango Delta (see Figure 9.1a) is a flood-driven, pulsing system with seasonal variations in discharge, fed with water annually from the Angolan highlands (Ramberg *et al.* 2006). Its surface area changes annually depending on the magnitude of hydrological inputs (Wolski *et al.* 2006). The flood waters arrive in the Panhandle portion of the Delta in December, peak in this region between April and May annually (Wolski and Murray-Hudson 2005; Wolski *et al.* 2006) and take several months to traverse the Delta's breadth, peaking as late as August in Maun. Therefore, there is a time lag between the arrival of the floods at the Panhandle and at different parts of the Delta (Wolski and Murray-Hudson 2005).

The Okavango Delta is the largest available freshwater body in Botswana (Merron 1993). It is found in the north-western part of the country in the Northwest District, commonly called Ngamiland (Figure 9.1). Most plausibly, it is an extension of the East African Rift Valley system (Gieske 1996). The Delta is a major tourist attraction in Botswana (Kolding 1996), and the main water source for north central Botswana (McCarthy 1992), in addition to supporting a small–scale commercial, subsistence, and recreational fishery (Merron and Bruton1995).

Commercial fisherman setting the fish net for the night. Fishing, both subsistence and commercial, is one of the important sources of livelihood in the Okavango Delta. Photo Credit: Alyne Delaney

The majority of the villages are in the Delta's panhandle. The hydrograph illustrates seasonal variations in mean discharge at Mohembo West hydrological station.

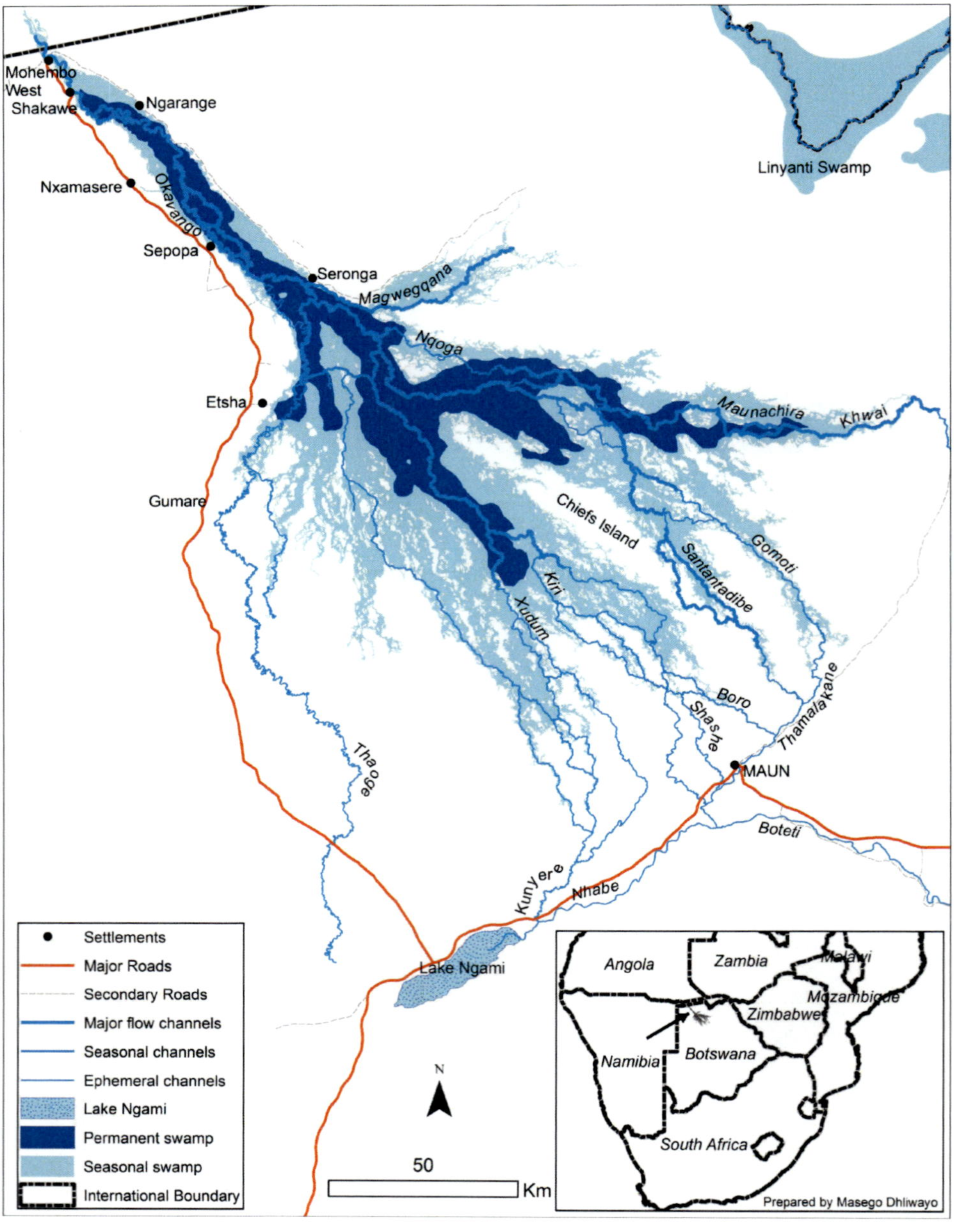

Figure 9.1: Map showing some of the key fishing villages around the Okavango Delta. Source: ORI GIS Laboratory.

Data collection

Fishery independent data: Experimental fish data used in this study are part of an ongoing long-term monitoring of the Delta's fish stocks where fish are sampled monthly, for 2 to 3 days per station at Ngarange and Seronga (see Figure 9.1). Standardised experimental nets, each made up of 9, 10m long panels of different mesh panels with a size range of 16–150 mm stretched mesh sizes, were used in experimental fishing (to sample the fish population). Nets were set for 12 hours overnight and 10 hours during the day to account for diurnal variations in fish movements. Each fish specimen was identified to species level and total length (TL) taken from the tip of the snout to the tip of the caudal fin (NB: for fish with forked tail fins, the tail fin was compressed and measurements to the longest tip of the compressed tail fin were taken). Fish maturity stages (for determination of spawning status) were also determined using a six-stage key where stage 1 was immature and stage 6 was spent fish.

Fishery dependent data: Catch and effort data are collected from fishers who record their daily fish catches from the several villages around the Delta as shown in Figure 9.1. The Fisheries Division field officers in these villages collect the data monthly and these are used to calculate total yield from the fishery. Furthermore, about 40 commercial fishers around the Delta's panhandle (where the majority of commercial fishers are found) recorded their catches (species caught, total length measurements for each fish species, mesh size used), and indicated whether a net was set overnight or used in drive fishing. These data were used to evaluate the nature of commercial fishing in the Delta. The data used was collected over a three-year period (1999–2002).

Hydrological data: These data are collected daily by the Department of Water Affairs at the Mohembo hydrological station (Figure 9.1) where the Okavango River enters the Delta in Botswana from Namibia (and Angola upstream).

Structured interviews: While Mosepele (2001) estimated that there were 40 commercial fishers in the Delta, and Bokhutlo *et al.* (2005) estimated about 50 commercial fishers, this study interviewed 30 'known' commercial fishers in the Delta's panhandle (about 70% of known commercial fishers) in 2008. A purposive sampling method was used to identify the commercial fishers and then a structured questionnaire was subsequently administered. The main purpose of the questionnaire was to evaluate the fishers' perceptions on the new fishing regulations.

Data analyses

Biological data: All fish data were captured in Pasgear (Kolding and Asmund 2006), a customised database package for managing and analysing fish data. Subsequently, fish biological indices – i.e. diversity, relative abundance (CPUE), spawning intensity (maturity stage 5), index of relative abundance (IRI), and mean length – were calculated in Pasgear (Kolding and Asmund 2006). Single factor

ANOVA was used to determine intra and inter annual variations in the biological variables where the level of significance was set at 95% confidence. Simple linear regression was also conducted to determine the nature and strength of the relationship between fish biological indices and mean discharge. DCA/DECORANA was used in this study to search for patterns in community shifts similar to Ibarra and Stewart (1989). Subsequently, a linear regression between Axis 1 scores of the DCA/ DECORANA analysis and discharge was done to determine the strength of the relationship.

Questionnaire data: These data were analysed in SPSS software and frequencies were subsequently calculated to estimate fisher perceptions on several management issues on the Delta's fishery.

Results and discussion

The fish and fishery dynamics

There are approximately 71 different fish species in the Delta (Ramberg *et al.* 2006), ranging from small cyprinid species (e.g. *Barbus fasciolatus*) to the largest species in the Delta (*Clarias gariepinus*) (Skelton 2001). Similar to most floodplain species (see Welcomme 1985; Lowe-McConnell 1987), the Delta's fish community structure undergoes significant temporal variations in composition (Mosepele *et al.* 2011) and relative abundance (Figure 9.3) driven by the seasonal flood pulse. Figure 9.2 reveals that seasonal flooding has a significant effect on the relative abundance of *Schilbe intermedius* (p= 0.00, df = 11, F = 31.21, r^2 = 0.76), *Clarias gariepinus* (p= 0.01, df = 11, F = 12.46, r^2 = 55), *Marcusenius altisambesi* (p= 0.01, df = 11, F = 10.78, r^2 = 0.52), and *Brycinus lateralis* (p = 0.02, df = 11, F = 6.96, r^2 = 0.41). Among these four species, *S. intermedius* had the best regression model (r^2 = 0.76), while *B. lateralis* had the poorest fitting model (r^2 = 0.41).

These seasonal changes, driven by the flood pulse, are caused by either longitudinal or lateral migrations (Merron and Bruton 1988) which result in fish 'concentration and dilution' effects, thereby affecting fish catchability of floodplain systems (Welcomme 1985). The fish species are distributed heterogeneously among the Delta's various habitats (Mosepele *et al.* 2009), where lagoons are key habitats for food/commercial fish species (Mosepele *et al.* 2011). The biology and ecology of the Delta's fish species is cued to the hydrological regime (Merron 1991; Mosepele 2008; Mosepele *et al.* 2009). Four of the key fish species in the Delta's fish community (Mosepele *et al.* 2011) undergo massive (peak) spawning 2 to 3 months before peak discharge. *M. altisambesi* (p= 0.01, F = 12.16, r^2 = 0.55) and *H. vittatus* (p= 0.01, F = 11.40, r^2 = 0.53) both undergo massive spawning three months before peak discharge (Figure 3), while *S. intermedius* (p= 0.001, F = 18.85, r^2 = 0.65) and *C. gariepinus* (p= 0.0495, F = 4.99, r^2 = 0.33) undergo massive spawning two months before peak discharge. This suggests that these species take advantage of available food in the temporally shallow inundated floodplains (by spawning before peak floods) which are characterised by high primary production, as observed by Mosepele (2008).

According to Merron (1993), the catfishes (*Clarias gariepinus* and *C. ngamensis*) undergo a seasonal feeding frenzy driven by the flood pulse where they feed voraciously on the bulldog (*Marcusenius altisambesi*). The silver catfish (*Schilbe intermedius*), undergoes seasonal changes in its feeding ecology where it actively feeds on terrestrial insects and mice which drown in rapidly advancing waters of the seasonal floodplains (Mosepele *et al.* 2012). This seasonal flood-driven feeding ecology was also observed by Siziba *et al.* (2011) who highlighted that juvenile fish production is regulated by micro-invertebrates, the abundance of which is driven by environmental variability.

Because of the nature of the fish species assemblage in this floodplain system, the Delta's fishery is characterised by diverse fishing gears (Mosepele *et al.* 2003; Mosepele 2008; Mosepele *et al.* 2009; Mmopelwa *et al.* 2009) and fishing methods (Mosepele 2000; Mosepele *et al.* 2007; Mmopelwa *et al.* 2009) to optimally exploit the different fish species in this system. Mosepele *et al.* (2003) showed that different fishers of different ages catch different fish species (or different sizes of similar species) in the Delta. Moreover, Mmopelwa *et al.* (2009) showed that mosquito nets harvested some of the smallest fish species (*Aplocheilichthys johnstoni*), fishing baskets harvested smaller sized cichlids (*Tilapia sparrmanii*) and caught the highest diversity of fish species, while traditional hook and line fishers caught larger sized cichlids (*Oreochromis andersonii*). Barrage traps caught smaller *Clarias* spp and the lowest diversity of fish species, while commercial fisher gill nets caught larger *Clarias* spp. Moreover, catch rates ($p = 0.01$, $df = 95$, $F = 6.94$, $r^2 = 0.07$) and total yield ($p = 0.01$, $df = 7$, $F = 17.56$, $r^2 = 0.75$) from the gill net fishery are driven by the flood regime (see Figure 9.4). Furthermore, mean monthly yield is approximately 25 tons (Figure 9.4), which translates into about 300 tons per annum from the gill net fishery over the 12-year period between 1996 and 2008.

Figure 9.4 also shows that fishing effort in the fishery is significantly driven by the seasonal flood regime ($p = 0.003$, $df = 9$, $F = 17.80$, $r^2 = 0.69$). This is perhaps due to the seasonal fishing methods that fishers practice as an adaptation to variations in fish catchability which is driven by environmental variability. Mosepele *et al.* (2007) showed that commercial fishers adopted 'drive fishing' methods as a coping strategy against seasonal variability in fish availability and also against net damage by hippos and crocodiles. This fishing method is used in lagoons between August and November when water levels are generally low in the Delta and the fish are 'concentrated' in deep lagoons. Therefore, to minimise net damage and optimise their fish catches within a short time, commercial fishers engaged in this fishing activity (see Table 9.1). Drive fishing was done in conjunction with seining where the fishers will then pull the net after they 'drove' the fish into it (Mosepele 2001).

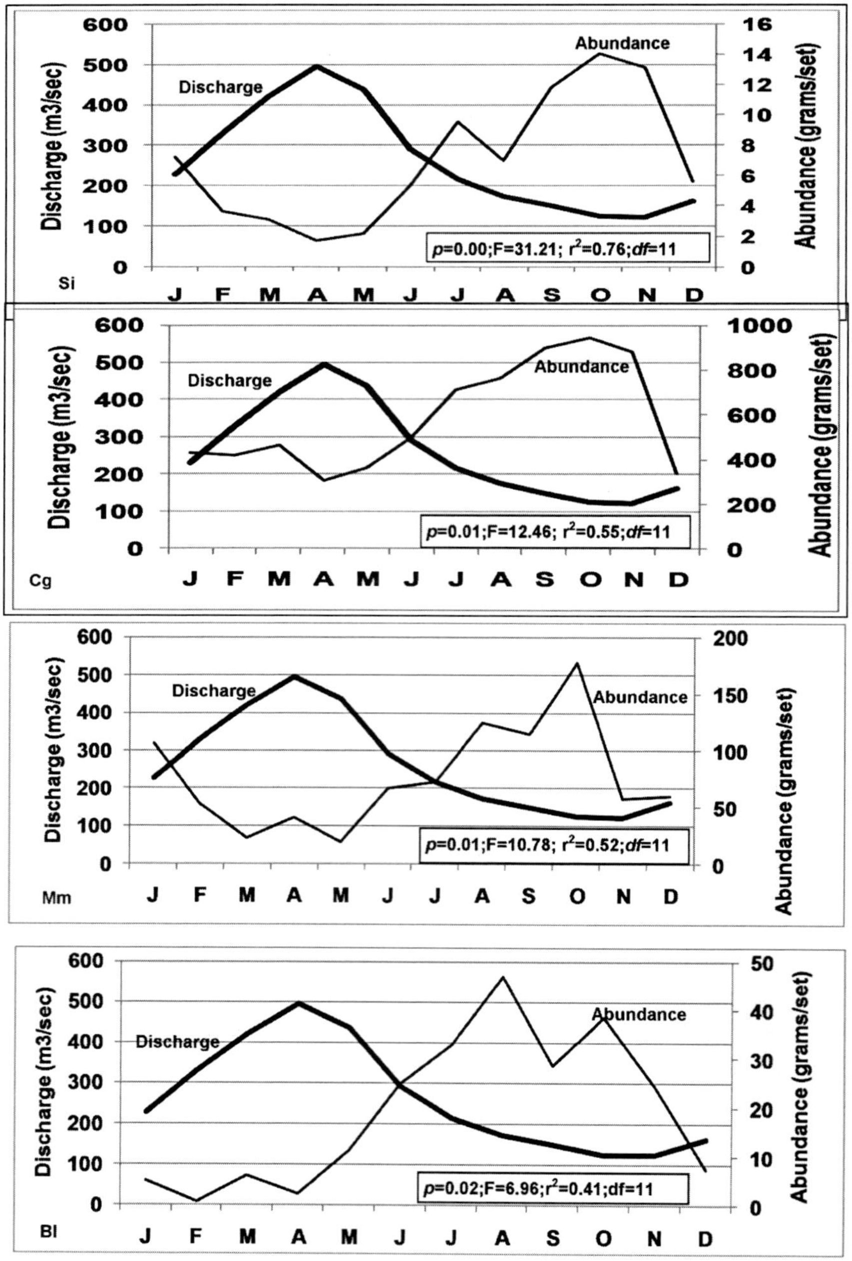

Figure 9.2 Seasonal variations in catch rates in relation to the flood regime of four selected fish species from the Delta's panhandle. Results from a simple linear regression are provided to show the nature and strength of the relationship. Note: Si = Schilbe intermedius; Cg = Clarias gariepinus; Ma = Marcusenius altisambesi; Bl = Brycinus lateralis.

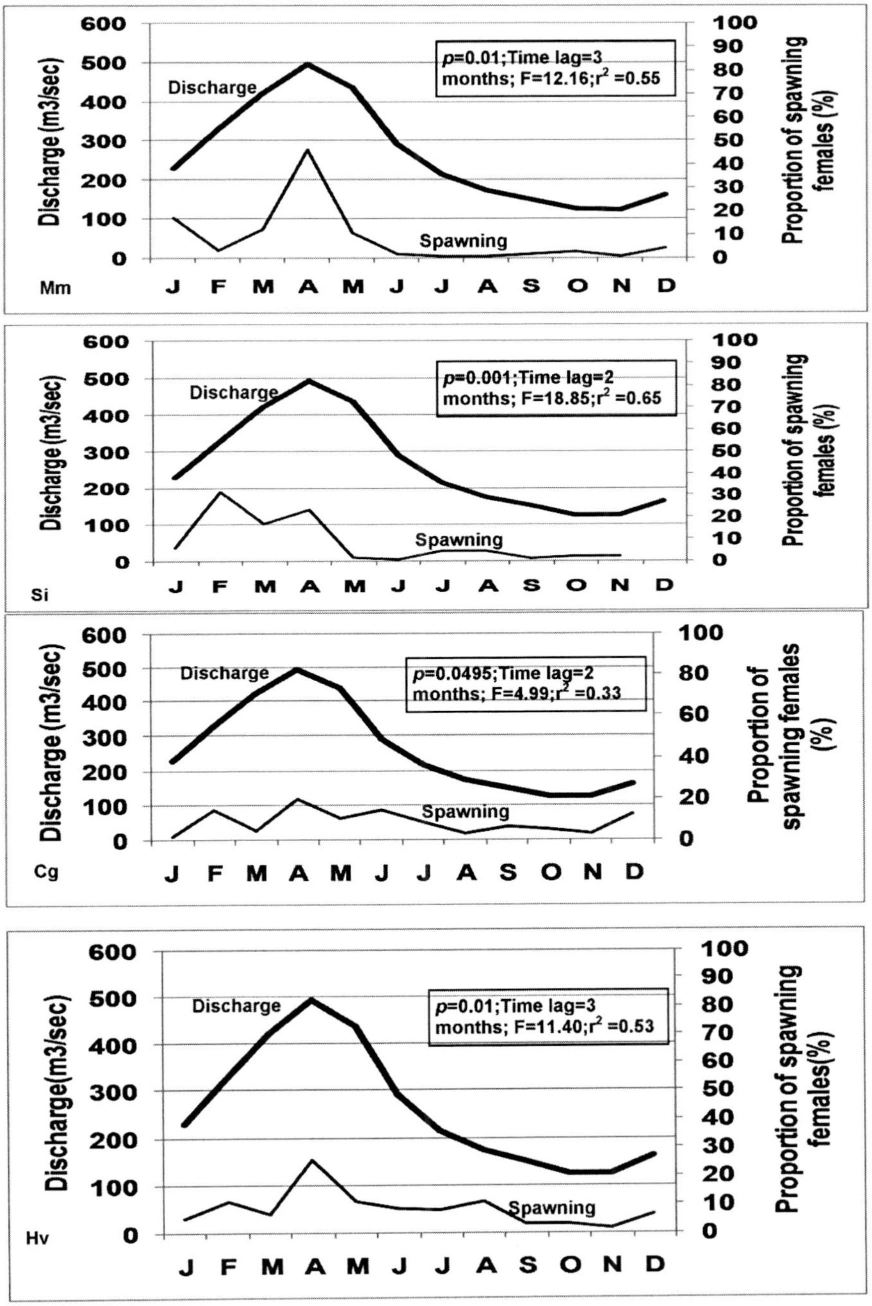

Figure 9.3: Effect of seasonal flooding on spawning of four selected fish species in the Delta. Note: Ma = Marcusenius altisambesi; Si = Schilbe intermedius; Cg = Clarias gariepinus; Hv = Hydrocynus vittatus

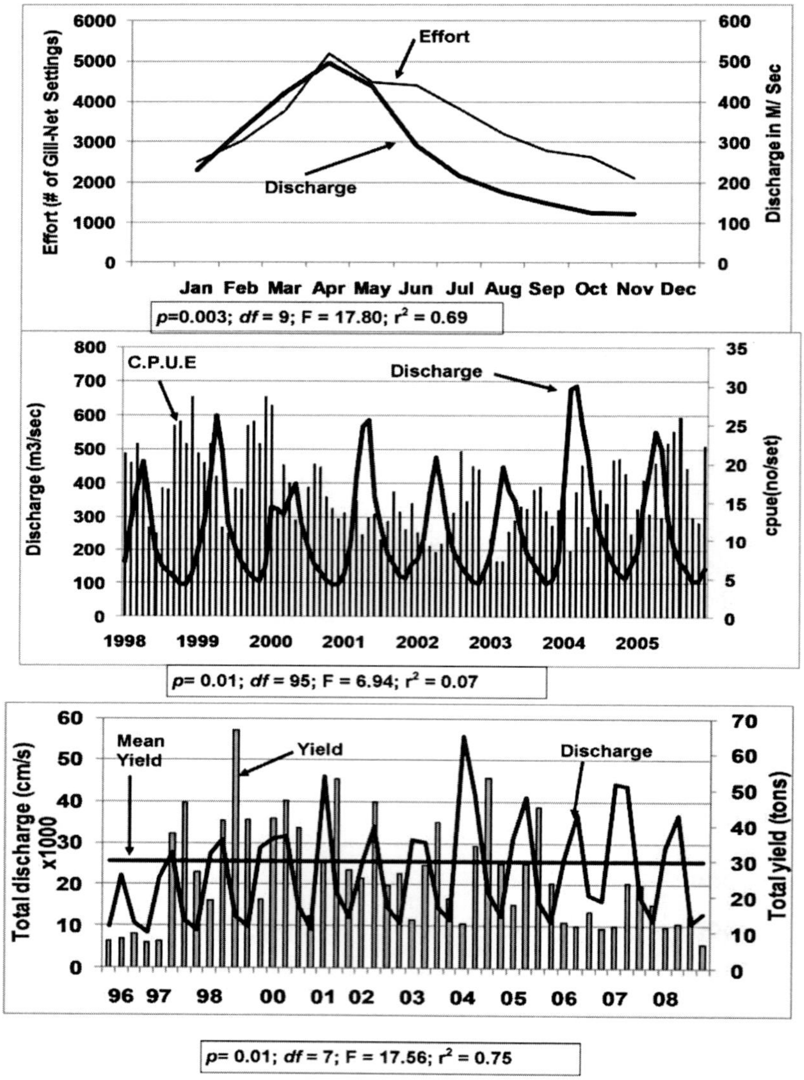

Figure 9.4: Effect of the flood regime on the Delta's fishery.

This form of fishing ensured an even distribution of fishing mortality among the three key commercial fish species (*O. andersonii, O. macrochir* and *Coptodon rendalli*) as shown in Table 9.1. This agrees with Bills (1996), who observed that seine catches had higher proportions of *C. rendalli* compared with overnight set gill nets. The majority of commercial fishers (88%) regard drive fishing as the most efficient fishing method (Table 9.2) and this confirms the observation of Mosepele *et al.* (2007) that catch rates from drive fishing are significantly higher than those from normal net setting. Furthermore, Mmopelwa *et al.* (2009) argue that different fishing gears are used at different times in different places as a coping strategy against spatio-temporal variations in fish availability. Therefore, gill nets and hook and line are generally used in relatively deep water (i.e. >1.5 m) while fishing baskets, barrage traps and mosquito nets are used in relatively shallow water (i.e. <1.5 m deep). Subsequently, the diverse fishing gears used in this fishery, together with the different fishing methods, resulted in what Jul-Larsen *et al.* (2003) call a *non-selective* fishing regime rationally exploiting a diverse fish community.

Table 9.1: Relative proportion of the three key commercial fish species in drive fishing catches and overnight fishing net set catches from the Okavango Delta commercial fishers.

Species	Overnight mean c.p.u.e	%N0	Drive mean c.p.u.e	%N0	P	F	Df
O. andersonii	39.06	51.50	104.30	36.80	0.00	37.92	18
O. macrochir	13.37	10.40	56.93	25.70	0.01	8.15	16
C. rendalli	10.74	9.90	44.06	22.20	0.00	22.9	16

Current management

There are between 2,700 (Bokhutlo *et al.* 2007) and 3,200 (Mosepele 2001) fishers in the Okavango Delta who use gill nets (factory made and traditionally made), fishing traps, fishing baskets and traditional hook and line (Mmopelwa *et al.* 2009) to exploit the Delta's fish species assemblage. Current management is based on classical input and output controls (Mosepele 2008) where certain fishing gear (e.g. mosquito nets) and some fishing methods (e.g. drive fishing) are prohibited (Botswana Government 2008). The Delta's fish resource is an important source of livelihoods for both the subsistence (Ngwenya and Mosepele 2008) and small-scale commercial fishers (Mosepele and Ngwenya 2010). It is also an important source of food security which assists the riverine communities as a 'mitigation resource' during times of food scarcity (Mosepele *et al.* 2006), where children from fishing households generally show a better nutritional status than those from non-fishing households (Nnyepi *et al.* 2007). The small-scale commercial fishing enterprises are a source of socio-economic empowerment for local fishers (Mmopelwa *et al.* 2005), where some fishers earn up to US$500.00/month (Mosepele and Ngwenya 2010) at current prices. In fact, Mosepele and Ngwenya (2010) highlight that commercial fishing contributes between 80–100% of household income for about 50% of commercial fishers in the Delta. This suggests therefore, that the Delta's fishery has several different use values (Figure 9.5), some of which

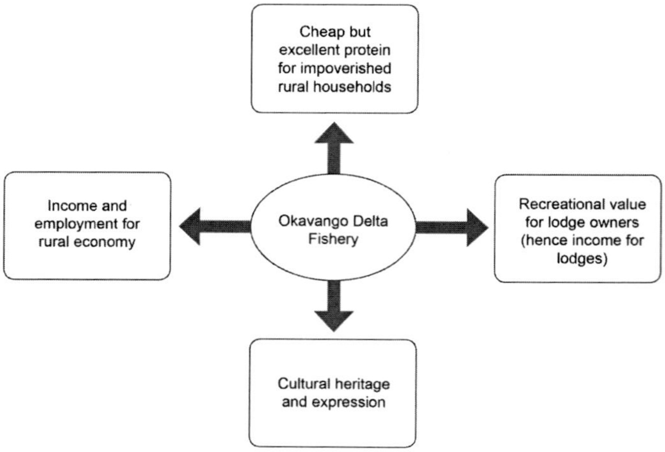

Figure 9.5: Different use values of the Delta's fishery.

are invariably in direct competition (e.g. commercial vs. recreational fishing). This direct competition has resulted in conflict in the fishery (Mosepele 2000).

As illustrated in Figure 9.5, the Delta's small-scale commercial fishery is a source of income and employment for the rural economy (Mosepele and Ngwenya 2010). Basket fishing (mostly by women) and the traditional hook and line fishing (mostly by young boys) (Mosepele *et al.* 2003) is a source of cheap and excellent protein for the household (Ngwenya and Mosepele 2008). The indigenous knowledge used by commercial fishers to target their preferred species (Mosepele *et al.* 2007) and the knowledge used in construction of traditional fishing gear (Mosepele 2001) suggest that the Delta's fishing is a source of 'tangible' and 'intangible' heritage. Mmopelwa (2006) has also shown that the fishery is a source of revenue for some fishing lodges. This situation – of multiple users, use of multiple gears, multiple use values, relatively open access and conflict in a dynamic environment – presents deep management challenges in this fishery.

While earlier research has shown that the Delta's fish stocks are not biologically over-exploited (Mosepele and Kolding 2003), concerns of over-exploitation (Mosepele 2000) resulted in the implementation of fish regulations in the Delta fishery. This management action was possibly driven by adoption of the FAO's precautionary approach to fisheries management, and resulted in the implementation of classical technical measures (e.g., mesh regulations, gear limitations, closed seasons, closed areas, etc.), input (licensing, state-regulated access, ownership, etc.) and output (e.g., quotas, size limits on fish harvested, etc.) controls in the fishery (Botswana Government 2008). The technical fishing regulations measures prohibit fishing with mosquito nets and also prescribe a fishing season. It needs to be noted, however, that the closed fishing season covers the period of peak fish spawning in the Delta's Panhandle (see Figure 9.3). In this respect there is a biological justification for the closed fishing season.

While the regulations do not include any specific mesh restrictions, commercial gill net fishers have been 'encouraged' to use larger mesh nets so that they harvest 'marketable sized' fish. The commonly used mesh sizes in the fishery are thus 100 to 125mm stretched mesh nets as observed by Mosepele (2000) and Bokhutlo *et al.* (2007). Input controls in the fishery include regulations that prohibit fishing without a fishing license (for commercial and recreational fishers). The fish regulations also confer private ownership of fish within Wildlife Management Areas (WMA), where the lessees are also given the 'power to search' any fisher who they find on 'their land' (Mosepele and Mosepele 2005). Output controls are enforced more on recreational fishers who are prohibited by law to catch and keep 'more than five fish per day' (Botswana Government 2008). These regulations are encouraging selective fishing of the larger sized species while the smaller sized fish species will remain relatively un-exploited.

66% of the (known) commercial gill net fishers in the Delta (Table 9.2) revealed that they had some misgivings about the nature and form of the fish regulations. Fishers exhibited innate indigenous knowledge of fish behaviour where the majority of fishers (83%) indicated that they fish at night because that is when fish are most active. This agrees with Mosepele *et al.* (2007) who showed that fishers use indigenous knowledge to target their preferred fish species. There was a 100% consensus among the fishers interviewed that the greatest negative impact of the fish regulations was that their nets will be damaged by crocodiles due to the prohibition on night (drive) fishing. Furthermore, 88% of the fishers indicated that drive fishing was an efficient fishing method, and prohibiting this method will invariably have a negative impact on their business. A high proportion of fishers (62%) reported that they disliked most prohibitions on night and drive fishing, and also the imposition of a fishing season. The majority of fishers (just over 40%) perceive damage to nets (by hippos and crocodiles), due to fishing regulation restrictions, and the lack of adequate compensation by Government as the biggest threats to the Delta's fishery.

Ecosystem changes

As previously stated, while there are no explicit mesh regulations in the fishery, the current management regime encourages selective fishing where fishing mortality is skewed towards larger sized species (e.g.*Hydrocynus vittatus, Clarias gariepinus, Serranochromis robustus*, etc) in the Delta. Moreover, prohibiting drive fishing skews fishing mortality towards *Oreochromis andersonii*, at the expense of the two other key commercial species, *O. macrochir* and *C. rendalli* (Table 9.1). Small-sized fish species are a key source of calcium, iron and zinc (Roos *et al.* 2007) and vitamin A (Roos *et al.* 2003) for Bangladeshi fishers. It is possible that

Table 9.2: Summary of proportions of commercial fisher perceptions on the new fish regulations and their potential impact on their fishing operations.

Question	Response/ Answer	%	Cumulative
1. Reason for fishing at night	i. Because fish are more active at night	83	
	ii. Protect nets from crocodiles	13	
	iii. Both a and b	4	100
2. How have the regulations affected you?	i. Negatively because the nets are being damaged by crocodiles	100	100
3. What do you dislike about the regulations?	i. Restriction of night fishin	31	
	ii. closed seaso	14	
	iii. Restriction of night fishin and fishing technique	17	
	iv. Other reason	38	100
4. Why did you drive fishing	i. Most efficient fishin metho	88	
	ii. Other reason	12	100
5. What do you think are the biggest threats to the Delta's fishery?	i. Crocodile and hippo damage of fishing gear and threats to fisher life	15	
	ii. Lack of government support in acquiring fishing gear and compensation for damaged gear	18	
	iii. Conflicts with the tourism sector	12	
	iv. Insufficient market for fish	6	
	v. Regulations ruin the fishing business	12	
	vi. No serious threats	21	
	vii. Ecosystem threats	12	
	viii. Others	4	100

small-sized fish species in the Delta (may) have the same nutritional value for poor fishing households. Therefore, prohibiting the use of mosquito nets suggests that these smaller sized species (e.g. *Aplocheilichthys johnstoni, A. katangae, Barbus afrovernayi,*etc.) will no longer be exploited, possibly negatively impacting nutrition in fisher households. Selective fishing in exploited populations causes system level changes in fish communities. Fishing regimes usually target large sized species (Christensen 1996; Law, 2000; Conover and Munch 2002; Badalamenti *et al.* 2002) like *H. vittatus* in the Delta due to mesh size regulations. In systems which evolved as top-down controlled (where predation pressure is the major factor regulating the fish community), predation pressure might be unwittingly removed/reduced (Friedlander and DeMartini 2002), whereupon the bottom-up processes might begin to dominate. This argument agrees with Essington *et al.* (2006) who highlight that the 'fishing down the food web' scenario in marine fisheries, due to selective fishing mortality that targets the top predators through mesh regulations, also results in a 'sequential replacement' of the top predators in the fish community by lower trophic level species. Alternatively, the target species might be prey for non-commercial species (Christensen 1996; Hall 1999) whereupon selective fishing may have a cascading effect on the fish community (Essington *et al.* 2006; Daskalov *et al.* 2007), due to a disproportionate increase in the predator group biomass because of selective fishing (mesh regulations). Subsequently, in those systems which evolved as bottom-up controlled, top-down processes may become more dominant due to the effects of selective fishing.

In fact, Harvey *et al.* (2006) suggest that this targeted fishing mortality might result in 'feedback' processes in the fish community where large sized prey, which were originally prey for the top predator, escape predation because their main predator is being selectively harvested. Consequently, the 'unexploited prey' may then start preying on the juvenile/ younger individuals of their main predator. Badalamenti *et al.* (2002) also discuss that predator/prey dynamics in exploited fish communities are changed significantly due to selective fishing mortality on top predators. Jackson *et al.* (2001) argue that some of the systematic negative changes observed in aquatic systems, such as disease outbreak, could be caused by a pervasive removal of higher trophic levels and not climatic changes as has been argued. Consequently, Jul-Larsen *et al.* (2003) advocate for a non-selective fishing regime in multi-species fisheries which harvest the fish community proportionally across all its trophic levels. This argument is premised on several fisheries in southern Africa where a non-selective fishing regime has not had any negative impact on the fish stocks under exploitation (Kolding *et al.* 1996; Songore 2002; Mosepele *et al.* 2003). Moreover, Mosepele *et al.* (2005) showed that the three main exploited species in the Okavango Delta fishery have different growth patterns from different parts of the Delta. Therefore, mesh size regulations in this system might effectively harvest one population (the *K* selected upper Delta stocks *vs.* the *r* selected lower Delta stocks), which might then drive an evolutionary change in this species towards the phenotype of the resultant unexploited stock.

Evolutionary changes

Prohibition of certain fishing gears (e.g. mosquito nets) and methods (e.g. drive fishing) encourages selective fishing in the Delta's fish community. Prohibition of drive fishing also encourages selective fishing on *O. andersonii* (Table 9.1). Selective fishing mortality (Conover and Munch 2002) causes evolutionary changes in targeted fish populations/stocks (Davis 1929; Law 2000; Walsh *et al.* 2006). Davis (1929) expresses concern that selectively harvesting 'marketable' (i.e. large sized fish) size classes of fish could inadvertently produce 'a slow growth rate of the species' under selective exploitation. Law (2000) argues that selective fishing (through mesh regulations) generates an evolutionary response in the exploited fish stocks which changes the sustainable yield. According to Conover and Munch (2002) and Walsh *et al.* (2006), mesh size regulations cause selective mortality on target fish stocks with a consequent unidirectional evolutionary change (i.e. a fishery exploited on mesh size regulations graduates towards 'genotypes with slower growth, earlier age at maturity, decreased fecundity' etc). This slower 'somatic' growth subsequently contributes towards overexploitaion of fish populations (Daskalov *et al.* 2007). Walsh *et al.* (2006) highlight that selective fishing mortality on a fish population actively selects for younger, smaller and 'slower growing phenotypes'. Basically this contradicts the dictates of the 'fathers' of modern fisheries science (e.g. Beverton and Holt 1957) whose yield-per-recruit models were predicated on selectively harvesting older, mature fish, while the younger fish were left as security for future yields. Indeed, while Schindler *et al.*'s (1998) bioenergetics model for Lake Victoria focuses more on determining catch levels for the Nile Tilapia at varying mesh sizes, they lost focus of the most fundamental issue of mesh size regulations – the evolutionary effects of selective mortality on Nile Tilapia as a consequence of mesh regulations. Essentially, Schindler *et al.* (1998) developed a management model for Lake Victoria based on a single species, with a specific focus on varying catch levels, but with the underlying motive of the MSY concept.

While Trippel (1995) makes an eloquent description of the evolutionary effects of selective fishing mortality (i.e. smaller fish, early maturing individuals, etc.), he falls short of attributing these to the effects of mesh size regulations. Conversely, Chen and Mello (1999) argue that 'size selective fishing mortality' due to mesh regulations is one major factor that drives exploited fish stocks to mature early at smaller sizes, or according to Harvey *et al.* (2006), selective fishing mortality affects the reproductive success of exploited stocks. Rochet (1998) showed that selective fishing causes phenotypic changes in exploited stocks on growth and reproductionwhich should be taken into account in stock assessments. According to Rochet (1998), failure to account for these phenotypic changes (which invalidates Beverton and Holt (1957), Ricker (1944) and other classical fisheries theorists) could result in over-estimating stock abundance based on the yield-per-recruit models. This may result in higher than optimum efforts to attain MSY and may indeed result in a stock collapse. This stock collapse is caused by loss of resilience in the exploited stocks by targeting large sized individuals with relatively low productivity and a correspondingly low sustainable biomass yield (Law *et al.* 2012).

Socio-economic impacts

Inland fisheries are a major source of livelihood and an important cultural activity for riparian communities living within their periphery (Cerdeira *et al.* 2000; Zwieten *et al.* 2003; Malasha 2003; Mosepele *et al.* 2006). Over time, these communities have evolved complex management paradigms aimed at ensuring sustained yields from their fish resource. Because of the multi-species nature of inland fisheries, inland fishers have therefore used a variety of fishing gears to exploit different trophic levels of the fish community (Mosepele *et al.* 2003; Zwieten *et al.* 2003). Fishers in the Okavango Delta use gill nets of different meshes (Mosepele 2003; Mosepele *et al.* 2003) to harvest the different species in the Okavango Delta fishery. The small-scale commercial fishery in the Okavango Delta harvests 31 different fish species, while the traditional fishery harvests 56 different fish species (Mosepele *et al.* 2003). It has been argued that mesh size regulations in the Delta will not only skew the fish community towards a preponderance of smaller sized species (Mosepele and Mosepele 2005; Mosepele *et al.* 2007a), but might also threaten household food security (Mosepele *et al.* 2007a). Moreover, Mosepele and Ngwenya (2008) have shown that fish is a major source of nutrition to HIV-positive fishers in the Delta. Mesh size regulations would remove this important source of nutrition for marginalised households. According to Malasha (2003), Zambian fishermen complained and refused to comply to mesh size regulations because 'they were losing many fish'. Interestingly, the authorities at the time (1949) felt that the fish species the fishers wanted to exploit "were not economically valuable and hence the mesh sizes could not be changed!" (Malasha 2003: 256).

The fisheries of the Lower Amazon are an important source of income and food for the regional economy (Almeida *et al.* 2001). According to Cerdeira *et al.* (2000), the approximately 2,000 fishermen of the Monte Allegre fishery in the Amazon use gill nets of different mesh sizes to harvest over 70 different fish species. The non-selective nature of the different gears used, and the gill nets of varying mesh sizes, effectively harvest the fish community across its different trophic levels. Furthermore, fishermen use gill nets of varying mesh sizes to harvest approximately 14 different fish species in Billings Reservoir, Brazil (Minte-Vera and Petrere 2000), while varying mesh sizes are used to exploit approximately 60 different fish species in the River Niger (Crespi 1998). It is therefore possible that implementing selective fishing mortality in these systems, while possibly good for the economic objectives of these fisheries (i.e. increasing yields for commercial operations), may ultimately deprive local communities of an important source of food. This argument is premised on the observation that subsistence fishing is generally the main source of food in these marginalised communities. Since the subsistence fishery is the most non-selective fishery using a wide range of mesh sizes to exploit food species (Mosepele *et al* 2003), these fishers will be hardest hit by mesh size regulations, to the detriment of their dependants.

Need for a paradigm shift?

The basic assumption behind the MSY concept is that a sustainable (*constant*) catch level can be maintained in a fishery (assuming a constant parameter system) indefinitely. This idea was then formalised through mesh size regulations to ensure that a fish stock is harvested at a level where according to Ricker (1944) mortality was equal to growth, to ensure maximum returns from the fishery. This approach was immortalised in fisheries management through Beverton and Holt's (1957) yield-per-recruit model and was subsequently transferred into tropical fisheries management by Pauly (1997, 1998). Despite earlier warnings that the MSY concept might not be the best management approach for mixed fisheries (Ricker 1954), this single-species model was assiduously integrated into contemporary fisheries management tools, to the detriment of world fish stocks. World fish stocks are now in complete decrease (Trippel 1995; Hall 1999; Hall and Mainprize 2004), ostensibly caused by mesh regulations implemented to achieve MSY from exploited stocks (based on the preceding discussion). The underlying philosophical construct behind this concept was that fish are an inexhaustible resource as long as you remove a portion that maximises your (economic) benefit from the fishery and leave the remainder to restock the wild populations. Unfortunately, one very important factor that was overlooked was the obvious fact that fish are biological organisms, subject to non-linear laws of nature. Their abundance is subject to natural variations (chaotic behaviour that is incongruent with classical fisheries theory which assumes equilibrium conditions), which will indubitably result in fluctuations in recruitment, in growth and in mortality rates. Invariably, this subsequently negates the 'Bible' (Pauly 1998) of fisheries science, Beverton and Holt's (1957) yield-per-recruit model.

Pauly (1999) puts the current fisheries management problem into perspective when he proposes that it is both an ecological and livelihood problem for SSF. However, the series of proposals that Pauly (1999) makes to ameliorate this 'global fisheries crisis' miss the point. On the livelihood issue, he proposes an introduction of interest (Individual Transferable Quotas) to make the fisheries private property, which he assumes to be a 'silver bullet' in addressing a perceived 'tragedy of the commons' scenario. On the ecological question he proposes the introduction of MPAs (Christensen 1996; Pauly 1999, 1997b; Conover and Munch 2002) to act as refuges and nursery sites for exploited stocks.

The possibility that global fish stocks are facing a crisis as a consequence of flawed philosophical constructs appears not to form part of mainstream thinking in contemporary fisheries management. Instead, new approaches towards fisheries management (based on classical fisheries philosophy) are proposed as a panacea for fisheries management ills. These approaches include the ecosystem approach (Hall and Mainprize 2004), multi-species models (Kirkwood 1982; Harvey *et al,* 2006), dynamic system models (Christensen 1996) demarcation of refuges or MPAs (Christensen 1996; Friedlander and DeMartini 2002). However, most major world fish stocks continue to decline. It is worth noting that despite all these management proposals, the underlying motive is the MSY concept, which is premised on selective mortality as a function of mesh size regulations.

Synthesis and conclusion

Fisheries science is based on strict, linear mathematical principles which have *attempted* to reduce the biological world into a series of 'logical mathematical and statistical formulae' to make it easier to handle and hence to manipulate. This numerical management of fisheries (Acheson and Wilson 1996) is predicated on what May (2003) discusses as the Newtonian philosophy, a construct which essentially argues that nature is explained/governed by mathematical laws. According to Fjeland (2001), this is based on Plato's theory of knowledge (i.e. 'geometry as the paradigm of knowledge'). However, the rigid geometric thinking of classical fisheries theorists is incongruent with the dynamism evident in floodplain fisheries. This is because fish populations are chaotic in nature (Acheson and Wilson 1996) and are not linear in their behaviour. Therefore, a classical approach to fisheries management in flood-pulsed tropical systems such as the Okavango Delta is incompatible with the dynamic nature of floodplain fisheries.

The traditional multi-gear fishery exploiting the Delta's multi-species assemblage is what Jul-Larsen *et al.* (2003) recommended for multi-species tropical assemblages that are characteristic of floodplain fisheries. Not only does this exploitation regime allow fishers to mitigate against fluctuations in fish availability due to environmental variability, but it also makes the fishery maintain its role as a foundation of cultural expression and heritage. Because of their rigid and mechanistic philosophical constructs, classical management regimes hamper fishers' ability to adapt to variations in fish availability. Management of these fisheries should not follow these mechanistic and linear constructs, but should rather be based on adaptive processes with inbuilt feedback systems. Managing fisheries for a 'non-selective' harvesting regime is the best approach towards a rational exploitation of fisheries (Law *et al.* 2012), instead of selective exploitation regimes which are incongruent to food security needs (Garcia *et al.* 2012) of SSF (floodplain) fisheries. This non-selective approach invariably allows for a multi-gear fishery as the nature of floodplain fisheries aims at exploiting various fish species within the fish community. Allowing various traditional fishing gears (methods) in these fisheries will enable them to maintain the role of 'melting pots' of (fishing) culture and tradition in the fishing sector. In fact, non-selective fishing regimes in SSF is close to ecological optimum exploitation (Kolding and van Zwieten 2011), which is an operationalisation of the ecosystem approach to fisheries (Garcia *et al.* 2012).

Ultimately, the real management challenge to inland SSF is to develop a basket of indicators to monitor these fisheries and to adopt a holistic approach to resolving the conflict that is inherent in these fisheries. This basket of indicators should include (but not be restricted to) mean length indices, mean catch rates (i.e. CPUE catch per unit effort), species diversity, and fishery response time to floods of previous years (Welcomme 1999), changes in fisher population size, and changes in mesh size use in the fishery. All these, however, highlight the need for enhanced data quality and collection from all facets of inland fisheries (i.e. gill net fishery, recreational fishery, traditional fishery).

Finally, the key question that needs to be asked is whether classical fisheries models are relevant in contemporary fisheries management paradigms? There is need to adopt a holistic approach to fisheries management, where the ecosystem needs are integrated into social needs of marginalised communities. There is need for a paradigm shift in fisheries management, where the chaotic behaviour of fish populations is integrated into adaptive management approaches that are fishery specific and not based on over-generalisations. A holistic and comprehensive management approach to fisheries may indeed see the fish resource used as a vehicle to achieve some of the Millennium Development Goals (MDGs) that are proving elusive to most African countries. Thus the basic framework of how SSF should be managed perhaps lies in the definition of the sustainable development framework.

References

Acheson, J.M and Wilson, J.A. (1996). "Order out of Chaos: The Case for Parametric Fisheries Management", *American Anthropologist*, New Series, 98(3), 579–594.

Almeida, O.T., McGrath, D.G and Ruffino, M.L. (2001). "The commercial fisheries of the lower Amazon: an economic analysis", *Fisheries Management and Ecology*, 8, 253–269.

Ames, T (2003.). *Putting fishermen's knowledge to work, the promise and pitfalls,* Fisheries Centre Research Report, University of British Columbia.

Badalamenti, F., Anna, G.D., Pinnegar, J.K., Polunin, N.V.C. (2002). "Size-related trophodynamic changes in three target fish species recovering from intensive trawling", *Marine Biology*, 141, 561–570.

Bas, C. (2000). "Fishery science: analysis and present situation", *Contributions to Science*, 1 (4), 489–510.

Beverton, R.J.H. and Holt, S.J. (1957). *On the Dynamics of Exploited Fish Populations,* Chapman and Hall.

Bokhuto, T, Kootsositse, M.V and Mosepele, K. (2007). "Okavango Deltafishery frame survey", Department of Wildlife and National Parks, Gaborone.http://www.scribd.com/doc/37412763/ODMP-Fishery-Frame-Survey-Draft-Report

Botswana Government. (2008). *Fish protection regulations*, Government Printing and Publishing Services, Gaborone, Botswana.

Bulletin Statistique. (1912). *Bulletin statistique des peches maritimes des pays du nord de L'Éurope*. Volume VI. Copenhagen, Denmark.

Cerdeira R.G.P and Ruffino, M.L. (2000). Fish catches among riverside communities around Lago Grande de Monte Alegre, Lower Amazon, *Fisheries Management and Ecology*, 7, 355–374.

Chen, Y and Mello, L.G.S. (1999). "Growth and maturation of cod (Gadus morhua) of different year classes in the Northwest Atlantic, NAFO subdivision 3Ps", *Fisheries Research*, 42, 87–101.

Christensen, V., (1996). "Managing fisheries involving predator species", *Reviews in Fish Biology and Fisheries,* 6, 417–442.

Conover, D.O and Munch, S.B. (2002b). "Sustaining Fisheries Yields Over Evolutionary Time Scales", *Science*, 297, 94–96.

Daskalov, G.M., Grishin, A.N., Rodionov, S. and Mihneva, V. (2007). "Trophic cascades triggered by overfishing reveal possible mechanisms of ecosystem regime shifts", *PNAS*, 104(25), 10518–10523.

Davis, F.M. (1929). "Preliminary note on experimental trawling with cod-end meshes of different sizes", *Journal du Conseil Interational por l'Exploration de la Mer,* IV (3), 287–299.

Davy, J. (1854). "Some Observations on the Ova of the Salmon, in Relation to the Distribution of Species", *Proceedings of the Royal Society of London*, Vol. 7, 362–363.

Essington, T.E., Beaudreau, A.H and Wiedenmann, J. (2006). "Fishing through marine food webs", *PNAS*, 103(9), 3171–3175.

FAO. (1955). *FAO Yearbook of Fishery Statistics* – Vol IV, Part 2, Rome, Italy.

Fjeland, R., 2001. "Facing the problem of uncertainty", *Journal of Agricultural and Environmental Ethics*, 15, 155–169.

Friedlander, A.M and DeMartini, E.E. (2002). "Contrasts in density, size, and biomass of reef fishes between the northwestern and the main Hawaiian islands: the effects of fishing down apex predators", *Mar Ecol Prog Ser*, 230, 253–264.

Garcia, S.M., Kolding, J., Rice, J., Rochet, M.-J., Zhou, S., Arimoto, T., Beyer, J.E., Borges, L., Bundy, A., Dunn, D., Fulton, E.A., Hall, M., Heino, M., Law, R., Makino, M., Rijnsdopr, A.D., Simard, F., and Smith, A.D.M. (2012). "Reconsidering the consequences of selective fisheries", *Science*, 335, 1045–1047.

Gieske, A. (1996). "Modeling surface outflow from the Okavango", *Botswana Notes and Records*, 28, 165–192.

Gislason, H and Rice, J. (1998). "Modelling the response of size and diversity spectra of fish assemblages to changes in exploitation", *ICES Journal of Marine Science*, 55, 362–370.

Graham, M. (1935). Modern theory of exploiting a fishery, and application to North Sea trawling, *Journal du Conseil Interational por l'Exploration de la Mer*, X(3), 264–274.

Hall, S.J and Mainprize, B. (2004). "Towards ecosystem-based fisheries management", *Fish and Fisheries*, 5, 1–20.

Hall, S.J. (1999). "Managing fisheries within ecosystems: can the role of reference points be expanded?"*Aquatic Conser6: Mar. Freshw. Ecosyst.* 9, 579–583.

Harvey, C.J., Tolimeri, N. and Levin, P.S. (2006). "Changes in body size, abundance, and energy allocation in rockfish assemblages of the Northeast Pacific", *Ecological Applications*, 16(4), 1502–1515.

Hilborn, R. and Walters, C.J. (1992). *Quantitative fisheriesstock assessment: Choice, dynamics and uncertainty*, Chapman and Hall Publishers, New York.

Hjort, J. (1926). "Fluctuations in the year classes of important food fishes", *Journal du Conseil International por l'Exploration de la Mer*, I (1), 5–38.

Jackson, J.B.C., Kirby, M.X., Berger, W.H., Bjorndal, K.A., Botsford, L.W., Bourge, B.J., Bradbury, R.H., Cooke, R., Erlandson, J., Estes, J.A., Hughes, T.P., Kidwell, S., Lange, C.B., Lenihan, H.S., Pandolfi, J.M., Peterson, C.H., Steneck, R.S., Tegner, M.J and Warner, R.R. (2001). "Historical overfishing and the recent collapse of coastal ecosystems", *Science*, 293 (5530), 629–637.

Johansen, A.C. (1927). "On the migrations of the herring", *Journal du Conseil International por l'Exploration de la Mer*, 2 (1), 1–27.

Jul Larsen E, Kolding J, Overa R, Nielsen J. R, and Zwieten, P.A.M. (2003). "Management, co-management or no management? Major dilemmas in southern African freshwater fisheries. Part 1: Synthesis Report", FAO Fisheries Technical Paper 426/1. Rome

Junk, W.J., and Wantzenm, K.M. (2004). "The flood pulse concept: New aspects, approaches and applications – an update", Proceedings of the Second International Symposium on the Management of Large Rivers for Fisheries. FAO and Mekong River Commission, FAO Regional Office for Asia and the Pacific, Bangkok, Thailand, pp117–147.

Kirkwood, G.P. (1982). "Simple models for multi-species fisheries", pp 83–98 In D. Pauly and G.I. Murphy (eds) *Theory and management of tropical fisheries*, ICLARM Conference Proceedings 9, 360 p. ICLARM, Manila, Philippines.

Kolding, J. (1996). *Feasibility study and appraisal of fish stock management plan in Okavango,* University of Bergen, Norway.

Kolding, J., and Asmund, A. (2006). Pasgear II, Version 2.3, University of Bergen, Norway.

Kolding, J., Ticheler, H.J. and Chanda, B. (1996). "Assessment of the Bangweulu Swamps Fisheries", WWF Bangweulu Wetlands Project, Final Report. 51p

Law, L., Plank, M.J., Kolding, J. (2012). "On balanced exploitation of marine ecosystems: results from dynamic size spectra", *ICES Journal of Marine Science*, 69(2011), 602–614.

Law, R., 2000. "Fishing, selection, and phenotypic evolution", *ICES Journal of Marine Science*, 57, 659–669.

Lowe-McConnell, R.H. (1987). *Ecological studiesin tropical fish communities,* Cambridge: Cambridge University Press.

Malasha, I. (2003). "Colonial and postcolonial fisheries regulations: The case of Zambia and Zimbabwe", pp 253–266, In Jul-Larsen, E., Kolding, J., Overa, R., Nielsen, J.R and P.A.M. Zwieten (eds.) *Management, co-managementor no management? Major dilemmas in southern African freshwater fisheries* 2 case studies, FAO Fisheries Technical Paper 426/2

May, R. (2003). "The best possible time to be alive: The logistic map", pp 28–45, In G. Farelo (ed.) *It must be beautiful. Great equations of modern science*,London: Granta Books.

McCarthy, T.S. (1992). "Physical and biological processes controlling the Okavango Delta – A review of recent research", *Botswana Notes and Records* (24), 57–86.

Merron, G.S. (1993). "Pack-hunting in two species of catfish, *Clariasgariepinus* and *C. ngamensis*, in the Okavango Delta, Botswana", *Journal of Fish Biology*, (43), 575–584.

Merron, G.S. and Bruton, M.N. (1988). "The ecology and management of the fishes of the Okavango Delta, Botswana, with special reference to the role of the seasonal floods", J.L.B. Smith Institute of Ichthyology, Grahamstown, South Africa, Investigational Report No. 29.

Merron, G.S. and Bruton, M.N. (1995). "Community ecology and conservation of the fishes of the Okavango Delta, Botswana", *Environmental Biology of Fishes*, 43,109–119.

Merron, G.S. (1991). "The ecology and management of the fishes of the Okavango DeltaBotswana, with particular reference to the role of the seasonal flood", PhD Thesis, Grahamstown, RSA: Rhodes University. 171 pp.

Minte-Vera, C.V and Petrere, M. (2000). "Artisanal fisheries in urban reservoirs: a case study from Brazil (Billings Reservoir, Sa˜o Paulo Metropolitan Region)", *Fisheries Management and Ecology*, 7, 537–549.

Mmopelwa, G. (2006)."The value of the Okavango Delta: A natural resource accounting approach", PhD Thesis, University of Pretoria, South Africa.

Mmopelwa, G., Mosepele, K., Mosepele, B., Moleele, N. and Ngwenya, B. (2009). "Environmental variability and the fishery dynamics of the Okavango DeltaBotswana: The case of subsistence fishing", *African Journal of Ecology,* 47 (Suppl. 1), 1–9.

Mmopelwa, G., Raletsatsi, S. and Mosepele, K. (2005). "Cost Benefit Analysis of Commercial Fishing in Shakawe, Ngamiland, Botswana", *Botswana Notes and Records*, 37, 11–21.

Mosepele, K. (2001). "Description of the Okavango Delta fishery", Technical Report to the Fisheries Section, Ministry of Agriculture, Gaborone, Botswana.

Mosepele, K and Kolding, J. (2003). "Fish Stock Assessment in the Okavango delta, Botswana – preliminary results from a length based analysis", pp 363–390, In Bernard, T., Mosepele, K. and Ramberg, L. (eds). *Environmental Monitoring of Tropical and Subtropical Wetlands,* University of Botswana, Maun and University of Florida, Gainesville, FLA

Mosepele, K and Mosepele, M. (2005). "Spatial and Temporal Variability in Fishery and Fish Community Structure in the Okavango Delta, Botswana: Implications towards fisheries management", *Botswana Notes and Records,* Vol. 37, 280–291.

Mosepele, K and Ngwenya, B. (2010). "Socio-economic survey of commercial fishing in the Okavango Delta, Botswana", Gaborone: Bay Publishing.

Mosepele, K, Mmopelwa, T.G, and Mosepele, B. (2003). "Characterization and monitoring of the Okavango delta artisanal fishery", pp 391–413, In Bernard, T., Mosepele, K. and Ramberg, L. (eds), op cit.

Mosepele, K. (2000). "Preliminary Length Based Stock Assessment of the Main Exploited Stocks of the Okavango Delta Fishery", MPhil Thesis, Department of Fisheries and Marine Biology, University of Bergen, Norway.

Mosepele, K. (2003). "Trends in Fisheries Development and Fish Utilization in the Okavango Delta." <http://ebookbrowse.com/mosepele-pdf-d171401846>

Mosepele, K. (2008). "Flood pulse in a subtropical floodplain fishery and the consequences for steady state management", pp 56–62, In O. Totolo (ed.). *Proceedings of the 2nd IASTED Conference on Water Resource Management; Science and technologyinnovation for sustainable development*. Calgary, Canada.

Mosepele, K., Mmopelwa, G, Mosepele, B and Kgathi, D.L. (2007). "Indigenous knowledge and fish utilization in the Okavango Delta, Botswana: Implications for food security", pp 292–302, In A. Ahmed (ed). *World Association for Sustainable Development; Managing knowledge, technologyand development in the era of informationrevolution*, Griffith University, Australia.

Mosepele, K., Mosepele, B., Bokhutlo, T and Amutenya, K. (2011). "Spatial variability in fish species assemblage and community structure in four subtropical lagoons of the Okavango delta, Botswana", *Physics and Chemistry of the Earth,* 36(14–15), 910–917.

Mosepele, K., Mosepele, B., Wolski, P and Kolding, J. (2012). "Dynamics of the feeding ecology of selected fish species from the Okavango Delta, Botswana", *ActaIchthyologicaetPiscatoria* (in press).

Mosepele, K., Moyle, P.B., Merron, G.S., Purkey, D and Mosepele, B. (2009). "Fish, floods and ecosystem engineers; interactions and conservation in the Okavango Delta, Botswana", *Bioscience*, 59 (1), 53–64.

Mosepele, K., Ngwenya, B.N. and Bernard, T. (2006)."Artisanal fishing and food security in the Okavango Delta, Botswana", pp 159–168, In A. Ahmed (ed). *World Sustainable Development Outlook: Global and local resources in achieving sustainable development*. Geneva: Interdescience Publishers.

Mosepele, K., Williams, L. and Mosepele, B. (2005). "Assessment of the feeding ecology of the Silver Catfish (*Schilbe intermedius*, Ruppel, 1832) in a seasonal floodplain of the Okavango Delta, Botswana", *Botswana Notes and Records*, 37, 208–217.

Needler, A.W.H. (1930). "The migrations of haddock and the inter-relationships of haddock populations in North American waters", Contributions to *Canadian Biology and Fisheries*, 6 (8–10), 243–313.

Ngwenya, B. and Mosepele, K. (2007). "HIV/ AIDS, artisanal fishing and food security in the Okavango Delta, Botswana", *Physics and Chemistry of the Earth,* 32, 1339–1349.

Ngwenya, B. and Mosepele, K. (2008). "Socio-economic survey of subsistence fishing in the Okavango Delta, Botswana", Gaborone: Bay Publishing.

Nielsen, J.R., Degnbol, P., Viswanathan, K.K., Ahmed, M., Hara, M., and Abdullah, N.M.R. (2004). "Fisheries co-management—an institutional innovation? Lessons from South East Asia and Southern Africa." *Marine Policy* 28: 151–160.

Nnyepi, M., Ngwenya, B and Mosepele, K. (2007). "Food (in)security and child nutrition in Ngamiland", pp 281–291, In A. Ahmed (ed).op cit.

Pauly, D. (1997a). "Small-scale fisheries in the tropics: marginality, marginalization and some implications for fisheries management," pp 40–49, In Pikitch, E.K., Huppert, D.D., and Sissenwine, M.P. (eds.) *Global trends: fisheries management*, American Fisheries Society Symposium 20, Bethesda, Maryland.

Pauly, D. (1997b). "Points of view: Putting fisheries management back in places", *Reviews in Fish Biology and Fisheries*,7, 125–127.

Pauly, D. (1998). "Beyond our original horizons: the tropicalization of Beverton and Holt", *Reviews in Fish Biology and Fisheries*, 8, 307–334.

Pauly, D. (1999)."Fisheries management: putting our future in places", pp 355–362, In Newell, D. and Ommer, R. (eds) *Fishingpeople, fishingplaces: issues in Canadian small-scale fisheries*, Toronto: University of Toronto Press.

Ramberg, L., Wolski, P. and Krah, M. (2006). "Water balance and infiltration in a seasonal floodplain in the Okavango Delta, Botswana", *Wetlands*, 26(3), 677–690.

Ricker, R.W. (1944). "Further notes on fishing mortality and effort", *Copeia*, 1, 23–44.

Ricker, W.E. (1954). "Stock and recruitment", *Journal Fisheries ResearchBoard of Canada*, 11(5), 559–623.

Ricker, W.E. (1958). "Maximum sustained yields from fluctuating environments and mixed stocks", *Journal Fisheries ResearchBoard of Canada*, 15(5), 991–1006.

Rochet, M.J. (1998). "Short-term effects of fishing on life history traits of fishes", *ICES Journal of Marine Science*, 55, 371–391.

Roos, N., Islam, M.M and Thilsted, S.H. (2003). "Small indigenous fish species in Bangladesh: Contribution to Vitamin A, Calcium and Iron intake", *J Nutr*, 133 (11 suppl 2): 4021s- 4026s.

Roos, N., Wahab, A., Hossain, M.A.R and Thilsted, S.H. (2007). "Linking human nutrition and fisheries: incorporating micronutrient-dense, small indigenous fish species in carp polyculture production in Bangladesh", *Food and Nutrition Bulletin*, 28(2, suppl): s280-s293.

Rounsefell, G.A. and Everhart, W.H. (1953). *Fishery science; its methods and applications,* New York: John Wiley and Sons.

Schaefer, M.B. (1943). "The theoretical relationship between fishing effort and mortality", *Copeia*, 2, 79–82.

Schindler, D.E., J.F. Kitchell and Ogutij-Ohwayo, R. (1998). "Ecological Consequences of Alternative Gill Net Fisheries for Nile Perch in Lake Victoria", *Conservation Biology*, 12(1), 56–64.

Siziba, N., Chimbari, M.J., Masundire, H., Mosepele, K. (2011). "Spatial and temporal variations of microinvertebrates across temporary floodplains of the lower Okavango Delta, Botswana", *Physics and Chemistry of the Earth,* 36 (14–15), 939–948.

Skelton, P.H. (2001). *A complete guide to freshwater fishes of southern Africa*, Cape Town: Struik Publishers.

Songore N. (2002). "Fish Diversity Development in Lake Kariba, Zimbabwe, from 1960 – 2001", MPhil Thesis, University of Bergen, Bergen, Norway

Sund, O. (1927). "Biological statistics of the stocks of fish."*Journal du Conseil International por l'Exploration de la Mer*, II (2), 161–169.

Trent, L and Hassler, W.W. (1968). "Gill Net Selection, Migration, Size and Age Composition, Sex Ratio, Harvest Efficiency, and Management of Striped Bass in the Roanoke River, North Carolina", *Chesapeake Science,* 9(4), 217–232.

Trippel, E.A. (1995). "Age at Maturity as a Stress Indicator in Fisheries", *BioScience*, 45 (11), 759–771.

Tweddle, D., Bills, R., van Der Waal, B., Skelton, P., Kolding, J., and Nengu, S. (2003). "Fish diversity and fisheries in the Okavango Delta, Botswana", pp 97–100, In Alonso, L.E., Nordin, L-A., (eds), *A Rapid assessment of the Aquatic Ecosystem of the Okavango Delta, Botswana: High Water Survey,* RAP Bulletin of Biological Assessment, 27.

Walsh,W.R., Munch, S.B., Chiba, S and Conover, D.O. (2006). "Maladaptive changes in multiple traits caused by fishing: impediments to population recovery", *Ecology Letters*, 9, 142–148.

Welcomme, R.L. (1999). "A review of a model for qualitative evaluation of exploitation levels in multi-species fisheries", *Fisheries Management and Ecology*, (6),1–19.

Welcomme, R.L. (1985). "River fisheries." FAO Fishery resources and environment division, FAO Fisheries Technical Paper, No. 262.

Welcomme, R.L. (1998). "Framework for the development and management of inland fisheries", *Fisheries Management and Ecology*, (5), 437–457.

Wolski, P. and M. Murray-Hudson. (2005). "Flooding dynamics in a large low-gradient alluvial fan, the Okavango Delta, Botswana, from analysis and interpretation of a 30 year hydrometric record", *Hydrology and Earth System Sciences* J1 – HESS, 10, 127–137.

Wolski, P., Savenije, H.H.G., Murray-Hudson, M., and Gumbricht, T. (2006). Modelling of the flooding in the Okavango Delta, Botswana, using hybrid reservoir-GIS model, *Journal of Hydrology*, 331, 68–72.

Zwieten P.A.M. van, Goudswaard, P.C., and Kapasa, C.K. (2003). "Mweru-Luapula is an open exit fishery where a highly dynamic population of fishers makes use of a resilient resource base", In E. Jul-Larsen, J. Kolding, R. Overå, J. Raakjær Nielsen and P.A.M.

van Zwieten (eds) *Management, co-managementor no-management?Major dilemmas in southern African freshwater fisheries*, FAO Fisheries Technical Paper 426/2. Rome, FAO.

Acknowledgements

While the majority of data used in this chapter were fully funded by the Fisheries Division in the Ministry of Wildlife, Environment and Tourism (Botswana Government) as part of its long-term monitoring protocol, some of the data were funded by the Okavango Research Institute (University of Botswana). My appreciation goes to Mr Masego Dhliwayo and Mrs Anastacia Makati of the Okavango Research Institute GIS lab who assisted with map production and figures for this Chapter.

CONCLUSION

10 Partnerships and the Strategy to 'Stop' Power Games in Okavango Delta Natural Resource Management and Governance

Alyne DELANEY and Lapologang MAGOLE

Abstract

As is demonstrated in this volume, failures in natural resource management are increasingly being viewed as a result of traditional, single-sector approaches to managing resources. Such fragmented approaches are unable to take into account the interrelatedness of both ecological and socio-economic dynamics of natural resource systems, without which successful management cannot take place. For this reason, the ecosystem approach "strives to balance diverse societal objectives, by taking into account the knowledge and uncertainties about biotic, abiotic and human components of ecosystems and their interactions and applying an integrated approach to fisheries [and other NR] within ecologically meaningful boundaries" (FAO 2003: 6, in FAO 2005). The increased use and acceptance of the ecosystem approach to natural resource management (NRM) has stemmed from growing appreciation of the interconnectedness of ecological and socio-economic and cultural systems and a realisation that it is the only path towards the seemingly elusive societal goal of *sustainability*.

Introduction

New cross-disciplinary scientific data, research and networks are needed to identify areas of sectoral interconnectedness, to articulate and implement the necessary research, and to translate the results into advice for policy makers. In this volume, the chapters take a first step towards doing just that: by showing the interconnectedness of resources in the Okavango Delta and the struggles around the management of these natural resources.

Historically, the processes of NRM in Botswana follow a collapsed communal management system, and are fraught with struggles involving serious resource use conflicts (Bolaane 2004; Magole and Magole 2009). Such challenges served as part of the impetus towards the development of collaborative management solutions (Magole and Delaney, this volume). One critical solution has been the introduction of the concept of sustainability – sustainable development, sustainable livelihoods, and sustainable ecological functions. Unfortunately, what is considered sustainable environmentally and what is considered sustainable socially appear to conflict at times, thus pitting one group (e.g., conservationists) against another (development groups). The challenge is to create a system which incorporates these varying types of sustainability and gives them equal weight, enabling them to work hand-in-hand.

In most cases, it is only once the social and environmental aspects of resource management are addressed together that successful management can take place. Thus, as Borrini-Feyerbend *et al.* (2004) highlighted, multiple and varied stakeholders must be taken into account. From this understanding comes increased awareness of the need for integrated systems of management such as integrated natural resources management (INRM), integrated water resources management (IWRM), adaptive management, co-management and sustainable land management (SLM).

This book

The chapters in this volume were compiled in order to present the current state of the commons and NRM in the Okavango Delta, focusing on work being undertaken for integrated management. As shown in previous chapters, the decades from the pre-colonial, colonial and post-independence period in Botswana have seen much change in resource management. It needs to be noted that while Botswana experienced a milder form of colonial rule (so-called protectorate), colonial influence was exerted through a system of indirect rule where Chiefs and conservation groups were used to set and enforce rules and regulations limiting use and access to natural resources, especially wildlife in the case of Ngamiland.

In Botswana as a whole, and Ngamiland in particular, the post-colonial period witnessed a more direct and significant shift in natural resources governance, with top-down governance structures beginning to manage resources centrally and sectorally, in addition to some continued use of indigenous commons systems. Over time the centralisation and disintegration of NRM continued, as depicted in the preceding chapters, and became rooted and solidified in a monstrous sectoral system, rather than the sustainability desired by modern Botswana society. Attempts to integrate NRM, particularly in the Okavango Delta region, have had little success, and the situation continues to frustrate resource users and managers alike.

Each of these chapters has value as a stand-alone case study account. Together they form a valuable body of literature highlighting the complexity of – and the resolve required for –successful integrated natural resource management. The first section of the volume focuses on case studies which address governance issues with NRM plans in general, and most particularly with stakeholder participation. The second section of the volume addresses general issues facing natural resource use and management, including internal (community capacity issues) and external pressures (policy and legislative frameworks).

Though the chapters and case studies present unique situations, there are connecting threads and similar themes found throughout the volume. The most important themes discussed include: a) *the nature of NRM governance*; b) *true stakeholder involvement*; c) *power relations; d) the transfer of knowledge and co-learning; and e) the involvement of interested outsiders.*

Governance

Stakeholder involvement, transparency, equity and knowledge are critical aspects or principles of governance in general and of natural resource management in particular (Borrini-Feyerabend *et al.* 2004; Gray 2005).

Governance is an ambiguous term, in much the same way as 'sustainability', with a variety of meanings and definitions depending upon who is using the term. Yet, it can be said to connote a "more informally-based, decentralized, shared, collective, and inclusive decision-making structure, with multiple layers of engagement" (Gray 2005: 2). Most scholars agree that governance is beyond mere government (Kooiman and Bavinck 2013); societies are governed by a variety of efforts, a variety of institutions, and at a variety of levels (Torfing *et al.* 2012). The key with these institutions and levels is the idea of *interaction* among them with new theories of interactive governance emerging (see, for example, Kooiman and Bavinck 2005). Thus, the term 'interactive governance' has been coined: "Governing interactions are exchanges between actors that contribute to the tackling of societal problems and opportunities" while "society is comprised of a large number of governing actors, who are constrained or enabled by their surroundings" (Kooiman and Bavinck 2013: 11). Kooiman and Bavinck further explain,

> Actors are any social unit possessing agency or power of action. This includes individuals, associations, firms, governmental agencies and international bodies. The surroundings are captured by the concept of structure. Structure refers to the social frameworks within which actors operate, including culture, law, politics and economics, but also natural conditions such as geography and ecosystems (ibid).

In looking at NRM in the Okavango Delta, we are investigating governance structures and principles, and at policies which form an integral part of governance. The research field of the anthropology of policy (e.g. Shore and Wright 1997) has emerged around this topic of governance (and the later governmentality). The anthropology of policy defines policy as:

> a type of power which works both on and through the agency and subjectivity of individuals as ethically free and rational subjects. It argues that policy has become a major institution of Western and international governance and therefore, the study of policies provide unique avenues to analysing wider issues of governance (Jacobsen 2014: 35).

As Shore and Wright note:

> Policies are most obviously political phenomena, yet it is a feature of policies that their political nature is disguised by the objective, neutral, legal-rational idioms in which they are portrayed. In this guise, policies appear to be mere instruments for promoting effi-

ciency and effectiveness. This masking of the political under the cloak of neutrality is a key feature of modern power (1997: 8).

Yet, there is a host of other factors which filter policy from the documents in which it is written to the effect, if any, it has on people and resources at the local level. Equal attention should be given to this second aspect: whether and why policy actually gets implemented and how different implementation and the results may be from what is stated in a policy document. As Magole and Delaney shared in the introduction to the volume, the struggles in implementing the Okavango Delta Management Plan were in fact the systemic struggle with *integration*.

Stakeholder involvement and legitimacy

True stakeholder involvement is believed to be the key to legitimacy (Chambers 1983; World Bank 2007). Another paradigm shift is the belief that participation in environmental decision-making is a democratic right (Reed 2008) for citizens. As Arnstein's "ladder of participation" (1969) shows, however, there are different levels of participation, with a continuum of increasing degree of participation by stakeholders from passive dissemination of information to active engagement (Reed 2008). Implicit in much of the literature surrounding stakeholder work is the idea that increased participation (e.g, higher rungs on the ladder) is preferable (Reed 2008); for example, inviting stakeholders to a meeting at the end of the production or implementation of a management plan as a simple 'checking the box' exercise is not considered legitimate participation. Actually hearing, considering, and weighing the needs of every stakeholder group involved is not only fair for citizens and stakeholders, but vital for the success of management. As the "communicative action" theory suggests (Habermas 1987), fair participation is required that represents the diversity of relevant stakeholders as well as ensures equality among them.

With stakeholder participation, there will also be a diversity of opinions as differing groups have differing interests (Magole and Delaney; Magole and Kgomotso, this volume); stakeholder groups may also vary in the extent of power and influence they hold or wield (see Magole and Magole, this volume; World Bank 2007; Reed, 2008). "Power inequalities within groups represent an equally important barrier to meaningful engagement. It is necessary to consider how inequalities in age, gender and background can be overcome to enable stakeholders to participate on a level playing field" (Reed 2008: 2422). If unequal power relations are not addressed from the outset, this may have increased negative consequences in the long run (see Magole and Kgomotso; Magole and Magole, this volume). There are complex relations between nature and society, after all, and a careful analysis of the forms of access and control over resources livelihoods is needed in order to understand the implications for environmental health and sustainability (Watts 2000).

Stakeholder trust in the management decisions may be increased if the process of decision-making and participation is viewed as being transparent by the participants (Richards *et al.* 2004). Furthermore, such transparency throughout the process, in addition to stakeholder involvement, may result in greater understanding and acceptance (Cassidy *et al.* this volume; UNEP 2005; Reed 2008); establishing trust among participating stakeholders may transform previously conflicting and competing relationships into cooperative ones (Stringer *et al.* 2006). In this volume, this potential is demonstrated by the application of the Dash Board governance analysis to the Tubu community (Cassidy *et al.*, this volume).

Power relations

One critical take-home lesson for work in the Okavango, Botswana, Southern Africa, and indeed resource management throughout the world, is the issue of power. Local stakeholders have increasingly been given a platform through which to take part in the planning processes. Yet, as demonstrated by the ODMP process (Magole and Kgomotso, this volume), at the end of the day, the efforts for greater overall stakeholder involvement are not reflected in the results or impact of the process, which often remains skewed in favour of particular groups or stakeholders. Typically in NRM processes local stakeholders lose out. There are multiple reasons for this: in some cases, locals simply are not organised and thus do not have their voice heard; in other cases, the existence of certain stakeholders is completely overlooked; and in yet others, governance institutions are not constituted to facilitate power sharing, which in turn lessens the power of stakeholders further down the chain. *One of the key objectives of integration in general, and stakeholder participation in particular, should be power (re)distribution.*

Natural resources management is inherently a political process dealing with political conflict over control and access to scarce resources. As Magole and Magole (this volume) showed, the power dynamics found in the operation of Gudigwa Camp as a socio-economic development project actually served to disempower the Gudigwa people. Development itself is full of power relations (Taylor 2000). In this case NGOS, such as ACORD and CI (Magole and Magole, this volume) provide technical assistance to the Gudigwa community, however, as Twyman (1997:197) argues, such interventions are coercive rather than co-operative, and reflect a wide-spread prevalence of coercive conservation masked by "community-based" approaches.

As a part of this reality, one must remember that ODMP Plans (i.e. Action Plans, Implementation Strategy and Monitoring and Evaluation Plans) do not change the status quo; they give neither decision-making power nor authority to primary stakeholders. In fact, the plan continues to disregard their traditional indigenous knowledge and Common Property Resource (CPR) management institutions as demonstrated through the Fish Protection Regulations of May 2008

(Mosepele, this volume). These regulations paint local fishing communities as a potential threat to the resource and place control measures on how the resource is accessed and used, especially by these local, small-scale commercial fishermen. The new regulations assume that the CPR mechanisms of the local communities have failed or are themselves a threat to the resource. As of yet, no powerful civil societystructures exist in Botswana in general and in Ngamiland in particular to represent the interests of these communities in decision-making about the Okavango Delta use and management.

Admittedly the Okavango Delta is a large and fragile system which has over time attracted multiple interests from multiple stakeholder and thus, really requires a planning authority with powers and resources to plan for a true integrated ecosystem management system, which guards against abuse of natural resources and the people. In the search for sustainability of the resources and livelihoods, the social pillar, in the form of socio-political issues of power, equity and social justice need to be resolved in order to afford innovative initiatives, such as IWRM the opportunity to succeed.

Knowledge transfer and co-learning

Since the 1980s (Johannes 1981; Chambers 1983) there has been a call for local stakeholders' knowledge and traditional resource usage to be included in planning processes and in management plans (see Magole and Magole; Mosepele, this volume). Local knowledge is two-fold; there is life as it is experienced on a day to day basis, and indigenous knowledge accrued to community members over time and passed on to younger generations through demonstration and experience (Motsumi *et al.* 2012; Magole and Magole, this volume). Inclusion of either of these two forms of knowledge into planning and decision-making processes empower the community as they feel relevant and able to contribute to modern NRM. Magole *et al.* (this volume) showed how a community was empowered with a natural resources monitoring system (MOMS) which enabled villagers to collect and analyse data on different aspects of their resource use. This data was then used to make informed decisions and negotiate with other stakeholders. The chapter shows how it also worked as a tool for conflict resolution in NRM; such collaboration and show of trust in the local stakeholders serves to further empower them (Reed 2008).

Local knowledge can be combined with scientific knowledge for a more comprehensive understanding of complex and dynamic natural systems and processes, such as found with the Okavango Delta (Kolawole *et al.* 2013). Many researchers have identified scientific analysis and information as essential to the participatory process (Reed 2008; Chase *et al.* 2004; Fischer and Young 2007). There is also growing evidence that the combination of local and scientific knowledge could empower local communities as they, for example, assume a role in monitoring and managing environmental changes (Reed 2008; Thomas and Twyman, 2004).

Involvement of interested outsiders

The involvement of interested outsiders, such as the University of Botswana's Okavango Research Institute (ORI) (as outlined by Magole and Demotts, this volume) is key to successful integrated NRM. ORI, then HOORC (see Magole and Demotts, this volume), proved they were committed to "getting it right". They worked hard to give local residents the fullest opportunity possible to share their thoughts, concerns, and criticisms. ORI's approach to the ODMP process was to conduct consultations through multiple iterations over time, rather than using a one-shot approach commonly used by consultants. Faculty collected thoughts from community members, elaborated on issues to help increase awareness, went away, discussed the input amongst themselves, synthesised feedback, returned to Delta communities and met with local leaders, and then went repeated the entire process. The opportunity for locals to speak with outsiders who were not a part of the government on resource issues was a key factor in establishing willingness among the community to participate in the process. Such a process created space in which to build trust among stakeholders, crucial in an area characterised by conflicts emanating from competition for natural resource access and use. However ORI's commitment to local community participation was not enough to fully empower communities for successful negotiation of stake and position in the Delta NR governance structures. Other more powerful stakeholder such as government officials had to share this commitment and believe in genuine community involvement in NRM. The entrenched culture of government departments as the 'police' of resources will take time to uproot, and it remains a challenge to convince government to appreciate communities as equal partners in resource management.

The 'foreign' actors in the case of ODMP, i.e donors (IUCN and German DED), may be said to have sponsored community empowerment in that they provided funding for all stakeholder consultations conducted by ORI and development of the communication strategy by the coordinating Department of Environmental Affairs). However the fact that both the process and product of such consultations and communication was placed in the hands of the government allowed for the status quo to remain; where the powerful get heard and the powerless remain unheard. The ideal thing, as Magole and Magole (this volume) propose, would have been to sponsor a community-based natural resource broker (CBNRB) or facilitator who is answerable to the community and whose key performance area is to successfully negotiate community stake and position in NR governance in the OD.

Along these same lines, it was also challenging for disempowered communities to take their rightful place in resource management; years of marginalisation in natural resources governance has resulted in a culture of complacency and dependence on government departments to solve local community problems. Communities thus learned over time that they were not valued in the process of NRM, and it will require an equally long (perhaps longer) time for them to (re)gain confidence to demand and assume their rightful place in NR governance of the OD.

Conclusion

As the chapters in this book have shown, the ODMP is a step in the right direction towards integrating social, economic, and environmental goals in a NRM strategy for the Okavango Delta area. The plan led to measurable success towards reaching its objective of producing a comprehensive INRM plan for the Okavango Delta Ramsar site. Nevertheless, the plan is currently facing problems with implementation, as was the case with the earlier ORPMP (Magole and Delaney, this volume).

Though an integrated plan, the ODMP was produced by independent sectoral departments and has to be implemented along the same lines. Consequently, the individuals responsible for components of the plan were answerable to their own sectors, and sometimes even a different Ministry from those of the Secretariat and agency coordinating the ODMP. This separation made coordination challenging as the ODMP Secretariat did not have supervisory powers over the then planning and now implementing staff. With such a fragmented institutional cultural system, it is not surprising that implementation turned out to be even more difficult than the planning process.

As stated in Magole and Delaney (this volume) the District governance is much more integrated due to the existence of institutions such as the District Development Committee (DDC) and District Land Use Planning Unit (DLUPU). While this is not a perfect approach, as the structures exclude resource users, the answer to achieving integration lies in 'killing' the power game and bringing NRM down from central government to the local (District) level and working out how resource users can be involved.

References

Bolaane, M. (2004). The Impact of Game Reserve Policy on the River BaSarwa / Bushmen of Botswana, *Social Policy and Administration*, Vol. 38, No. 4, pp 399–417.

Borrini-Feyerabend, G., Pimbert, M., Taghi Farvar, M., Kothari, A., and Renard, Y. (2004). *Sharing power, learningby doing in co-managementof natural resources throughout the world*. Cambridge: IUCN.

Chambers, R. (1983). *Rural Development: Putting the Last First,* London: Longman.

Chase, L.C., Decker, D.J., and Lauber, T.B. (2004). "Public participation in wildlife management: What do stakeholders want?" *Society and Natural Resources* 17, 629–639.

FAO (2003). Fisheries Department. "The ecosystem approach to fisheries", FAO Technical Guidelines for Responsible Fisheries, No.4, Suppl. 2. Rome, FAO, 2003.

FAO (2005). Fisheries and Aquaculture topics. The ecosystem approach to fisheries management. Topics Fact Sheets, Text by S.M. Garcia and K.L. Cochrane, In: *FAO Fish-*

eries and Aquaculture Department [online]. Rome. Updated 27 May 2005. [accessed 28 March 2014]. http://www.fao.org/fishery/topic/13261/en.

Fischer, A., and Young, J.C. (2007). "Understanding mental constructs of biodiversity: Implications for biodiversity management and conservation", *Biological Conservation* 136, 271–282.

Habermass, J. (1987). *Theory of Communicative Action,* Cambridge: Polity Press.

Gray, T. (2005). *Participation in Fisheries Governance.* (ed), Springer, Dordrecht, Netherlands.

GIIRS n.d. "Emerging Market Assessment Resource Guide: Stakeholder Engagement" Pdf (on file).

Jacobsen, R.B. (2014). Power and Participation in Greenlandic Fisheries Governance The becoming of problems, selves and others in the everyday politics of meaning, Doctoral Thesis, Aalborg University, Aalborg, DK.

Johannes, R.E. (1981). *Words of the Lagoon: Fishingand Marine Lore in the Palau District of Micronesia,* Berkeley: University of California Press.

Kolawole, O.D., Mogobe, O. and Magole, L. (2013). "Political economy of integrated soil fertility management in the Okavango Delta, Botswana", Proceedings of the ICECESS 2013: International Conference on Environmental, Cultural, Economic and Social Sustainability held during November 14–15, 2013 in Venice, Italy,*World Academy of Science, Engineering and Technology*, Issue 83, pp. 512–520.

Kooiman, J. and Bavinck, M. (2005). The governance perspective, pp 11–25, In J. Kooiman, M. Bavinck, S. Jentoft, R. Pullin (eds) *Fish for life: Interactive governance for fisheries*, Amsterdam: Amsterdam University Press.

Kooiman, J. and Bavinck, M. (2013). "Theorizing Governability – The Interactive Governance Perspective", In M. Bavinck, R. Chuenpagdee, S. Jentoft and J. Koiman *Governability of Fisheries and Aquaculture; Theory and Applications,* MARE Publication Series 7: Springer Press.

Motsumi, S, Magole, L, Kgathi, D.L. (2012). "Indigenous knowledge and land use policy: Implications for livelihoods of flood recession farming communities in the Okavango Delta, Botswana", *Physics and Chemistry of the Earth,* 50–52 (2012) 185–195.

Magole, L. and Magole, I.L. (2009). "The Okavango: Whose Delta is it?" *Physics and Chemistry of the Earth,* 34 (2009) 874–880.

Reed, M.S. (2008). "Stakeholder participation for environmental management: A literature review", *Biological Conservation,* Vol. 141, pp 2417–2431.

Reed, M.S., and Dougill, A.J. (2002). "Participatory selection process for indicators of rangeland condition in the Kalahari", *The Geographical Journal* 168, 224–234.

Richards, C., Blackstock, K.L, and Carter, C.E. (2004). "Practical Approaches to Participation", SERG Policy Brief No. 1.Macauley.Land Use Research Institute, Aberdeen.

Robbins, P. (2004). *Political Ecology: A Critical Introduction*, Malden, Oxford and Carlton: Blackwell.

Shore, C. and Wright, S. (eds) (1997). *Anthropology of Policy: Critical Perspectives on governanceand power,* London: Routledge.

Stringer, L.C., Prell, C., Reed, M.S., Hubacek, K., Fraser, E.D.G., Dougill, A.J. (2006). "Unpacking 'participation' in the adaptive management of socio-ecological systems: a critical review," *Ecology and Society* 11, 39 (online).

SustainAbility (2007). "Practices and Principles for Successful Stakeholder Engagement, October 2007.http://www.sustainability.com/library/successful-stakeholder-engagement (accessed 29 march 2014).

Torfing, J., B.G. Peters, J. Pierre, and Sørensen, E. (eds). (2012). *Interactive governanceadvancing the paradigm,* London: Routledge.

Thomas, D.S.G., Twyman, C. (2004). "Good or bad rangeland? Hybrid knowledge, science, and local understandings of vegetation dynamics in the Kalahari", *Land Degradation and Development* 15, 215–231.

Twyman, C. (1997). "Community Development and Wildlife Management: Opportunity and Diversity in Kalahari Wildlife Management Areas, Botswana", PhD thesis, Department of Geography, University of Sheffield, UK.

UNEP. (2005). United Nations Environment Programme, "The Stakeholder Engagement Manual: The Guide to Practitioners' Perspectives on Stakeholder Engagement (Vol. 1), prepared by Stakeholder Research Associates Canada, July 2005.

Watts, M. (2000). *Political Ecology,* In E. Sheppard and T.J. Barnes (eds) *A Companion to Economic Geography,* Blackwell Publishing, Malden.

World Bank. (2007). International Finance Corporation (IFC), "Stakeholder Engagement: A good practice handbook for companies doing business in emerging markets", New York: New York.

Index

Defragmenting African Resource Management (DARMA)

edited by Dr. Mafaniso Hara (Institute for Poverty, Land and Agrarian Studies, University of the Western Cape) and Prof. Dr. Tobias Haller (Universität Bern)

Steve Donda; Mafaniso Hara; Maxon Ngochera; Erling Berge (Eds.)
Fragmentation of Resource Management on the South East Arm of Lake Malawi
Dynamics around Fisheries
The Southeast Arm of Lake Malawi and its catchment provide interdependent benefits and livelihoods to users, in particular the poor, who draw upon on the various resources (fish, land, water, forests, wildlife, etc.) in the area. Unfortunately, management of these resources remains fragmented, resulting in suboptimal social, economic and ecological outcomes. This book is an attempt to demonstrate the value of an inter-disciplinary approach in management of the South East Arm. It is one of seven case studies under the DARMA (Defragmenting African Resource Management in Southern Africa) project.
Bd. 3, 2014, 192 S., 29,90 €, br., ISBN 978-3-643-90505-5

Lindah Mhlanga; Kefasi Nyikahadzoi; Tobias Haller (Eds.)
Fragmentation of Natural Resources Management
Experiences from Lake Kariba
This book is an output of the Defragmenting African Resources Management (DARMA) Project. Lake Kariba and its environs are a complex transboundary ecosystem with multiple, competing and often contested resource uses. The individual chapters in the book describe the current sectoral approach to natural resource management, the interconnections, and the need to adopt an ecosystem approach. The book is an essential reading for students, researchers, academics and managers in the environmental and natural resources sectors interested in advancing the ecosystems approach in management of commons in general and Southern Africa in particular.
Bd. 2, 2014, 184 S., 29,90 €, br., ISBN 978-3-643-90400-3

Peter Mvula; Meya Kalindekafe; Paul Kishindo; Erling Berge; Friday Njaya (Eds.)
Towards Defragmenting the Management System of Lake Chilwa Basin, Malawi
The book presents results of the Defragmenting African Resource Management (DARMA) Project covering Lake Chilwa basin in Malawi.
The central theme is that in order to ensure resource base sustainability, research and management within the basin should adopt an ecosystems approach. Presently, research and management of the basin is sector-based, hence resource user conflicts are increasing. User demand for various resources is increasing rapidly mainly due to population increase and lack of alternative economic activities thereby presenting challenges to sustainable resource management. Specific areas of sectoral interconnections are highlighted and defragmentation options suggested.
Bd. 1, 2014, 144 S., 29,90 €, br., ISBN 978-3-643-90398-3

LIT Verlag Berlin – Münster – Wien – Zürich – London

Distribution: see CIP page